Agricultural Marketing

Agricultural Marketing

Editor: Albert Maxwell

www.callistoreference.com

Callisto Reference,
118-35 Queens Blvd., Suite 400,
Forest Hills, NY 11375, USA

Visit us on the World Wide Web at:
www.callistoreference.com

ISBN: 978-1-64116-295-1 (Hardback)

Trademark Notice: Registered trademark of products or corporate names are used only for explanation and identification without intent to infringe.

Cataloging-in-Publication Data

Agricultural marketing / edited by Albert Maxwell.
    p. cm.
Includes bibliographical references and index.
ISBN 978-1-64116-295-1
    1. Farm produce--Marketing. 2. Produce trade. 3. Agriculture--Economic aspects. 4. Marketing. I. Maxwell, Albert.
HD9000.5 .A37 2020
338.19--dc23

Table of Contents

Preface

The main aim of this book is to educate learners and enhance their research focus by presenting diverse topics covering this vast field. This is an advanced book which compiles significant studies by distinguished experts in the area of analysis. This book addresses successive solutions to the challenges arising in the area of application, along with it; the book provides scope for future developments.

Agricultural marketing deals with the services involved in the movement of agricultural products from the farm to the consumer. It is concerned with the planning, organizing, directing and handling of agricultural products to satisfy the farmer, producer and consumer. Agricultural marketing consists of various activities and services such as production planning, growing, harvesting, grading and packing as well as transporting, storage, food-processing and distribution of the products. It also includes the advertising and sale of agricultural produce. It provides market information to help direct these services. Modern agricultural marketing focuses on developing new marketing links between agribusiness, large retailers and farmers, through contract farming, group marketing and other collective actions. This book provides comprehensive insights into the field of agricultural marketing. It presents researches and studies performed by experts across the globe. It will serve as a reference to a broad spectrum of readers.

It was a great honour to edit this book, though there were challenges, as it involved a lot of communication and networking between me and the editorial team. However, the end result was this all-inclusive book covering diverse themes in the field.

Finally, it is important to acknowledge the efforts of the contributors for their excellent chapters, through which a wide variety of issues have been addressed. I would also like to thank my colleagues for their valuable feedback during the making of this book.

Editor

A milk marketing system for pastoralists of Kilosa district in Tanzania: market access, opportunities and prospects

Abel Leonard[1*], D.M. Gabagambi[1], E. K. Batamuzi[2], E.D. Karimuribo[2], R.M. Wambura[3]

[1]Department of Agricultural Economics and Agribusiness, Sokoine University of Agriculture, P.O. Box 3007 Morogoro, Tanzania.
[2]Faculty of Veterinary Medicine, Sokoine University of Agriculture, P.O. Box 3015 Morogoro, Tanzania.
[3]Institute of Continuing Education, Sokoine University of Agriculture, P.O. Box 3044 Morogoro, Tanzania.

Despite a large diversity of livestock species in Tanzania, most livestock keepers are not commercial oriented. However, this paper analyzed commercial settings particularly the efficiency of a pastoral milk marketing system in Kilosa District and identifies strategies to improve milk marketing. A cross-sectional research design was employed to collect data using interview schedule, focus group discussions and key informant interviews. Marketing margins, descriptive statistics and Structure-Conduct-Performance (SCP) model were used as analytical tools. The findings show that pastoral milk marketing is not efficient but profitable. This signifies potential for increasing income for pastoralist if milk efficiency was improved. Net profit margins realized per liter were TZS 332.00(1 USD = TZS 1800.00), TZS 65.00 and TZS 141.00 by producers, small scale milk vendors and retailers, respectively. The market information flow was not transparent. The pastoral system had no defined standards, grades, or product differentiation in packaging thus signaling market inefficiency. Commercialization focusing on promotion of pastoral milk marketing, through introduction of community based extension services and village community banks, empowerment of small and medium scale processors, training on proper milking and milk handling need to be considered in the future intervention strategies.

Key words: Marketing chains, profit margins, pastoralists, Kilosa district.

INTRODUCTION

The majority of the world's estimated 1.3 billion poor people live in developing countries where they depend directly or indirectly on livestock for their livelihoods (World Bank, 2008 and FAO, 2009). In Tanzania, out of the 4.9 million agricultural households about 36% are keeping livestock of whom 35% are engaged in both crop and livestock production (Njombe et al, 2011). According to Tanzanian Economic Survey of 2010, the sector grew by 3.4%, and contributed about 3.8% to the Gross Domestic Product (GDP) of which 40% came from beef, 30% from dairy and the remaining 30% from other livestock commodities.

*Corresponding author: Mr. Abel Leonard Department of Agricultural Economics and Agribusiness, Sokoine University of Agriculture, P.O. Box 3007 Morogoro, Tanzania. Email: abell82@yahoo.com

The main dairy animal in Tanzania is cattle which are classified as dairy for those that average about 2000 liters per lactation and dual purpose indigenous cattle producing around 300 - 500 liters and are mostly used for beef which are the majority (Njombe et al, 2011). The dairy cattle are kept by smallholder farmers and a few medium and large scale farmers. The indigenous cattle are kept by traditional livestock keepers in the pastoral and agro pastoral systems. The bulk of milk produced in the country originates from traditional cattle and is consumed at the household level with very little reaching the market. About 70% of the annually produced milk comes from the indigenous cows, whereas the dairy cows produce about 30% (Njombe and Msanga, 2008). Moreover, much of milk marketed comes from improved dairy cattle, which are located in urban and peri-urban centers while traditional cattle are located in rural areas (URT, 2005). Better cash income to the producers creates a production incentive, however, increased income could only be ensured through prevalence of an efficient marketing system (Mari, 2009).

Options for pastoralists to secure their livelihoods through current policy, legal and economic issues are encountered with many challenges and opportunities that need to be addressed in future efforts (Tenga et al, 2008) hence this study was intended to analyze and document the efficiency of the pastoral milk marketing system in Kilosa district so as to recommend the strategies for sufficient marketing. Albeit of different studies; Mung'ong'o and Mwamfupe (2003) documented the plight of the Maasai pastoralists who have moved to Morogoro and Kilosa districts as a result of socio-economic developments and environmental changes in Maasai land. It has been revealed that genuine social change is taking place among migrant Maasai and they are taking up several non-pastoral economic activities as alternative ways of earning their livelihoods. Mdoe and Mnenwa (2007) assessed options for pastoralists to secure their livelihoods through assessing the total economic value of pastoralism using total economic value (TEV) concept to capture all the economic values for manmade capital assets and the natural resources and found that pastoralism has been receiving little attention from the Tanzanian government in its development agenda due to under-valuation. The present study aimed at identifying the problems, obstacles, and other inefficiencies in the prevalent milk marketing system and explored the possibility for the solution of significant problems that might lead to increase in the physical and economic efficiency of pastoral milk marketing in Kilosa district.

MATERIALS AND METHODS

This study was carried out in Kilosa District, one of the six

Districts that comprise Morogoro Region. It adopted a cross section research design with purposive (aiming to pastoralists engaged in milk production) and simple random sampling. Qualitative and quantitative data were collected during as survey of 219 pastoralists selected randomly from pastoral communities. Data from interview schedule were complemented by information collected from 17 milk vendors, 31 retailers and 1 transporter using semi structured interview schedule, checklists and focus group discussions.

The Gross Margin (GM) analysis was carried out for different actors (pastoralists, small-scale milk vendors and retailers) in the value chain to determine the profitability of milk marketing. The structure, conduct and performance (S-C-P) model from the theory of industrial organization was employed to examine the causal relationship between market structure, conduct and performance of the milk marketing system in the study area as a measure of efficiency(Shaik et al,2005). The S-C-P approach is appropriate for identifying factors that determine the competitiveness of the marketing system, behavior of actors, and the success of milk industry in meeting performance goals. Buyer's concentration index, break even analysis and marketing efficiency ratio were also used.

Concentration ratio (CR_4) for milk vendors was also found. Kohls and Uhl (1985) suggested as a rule of thumb, four largest enterprises concentration ratio of 50% or more as an indication of a strongly oligopolistic industry and CR_4 between 25% and 50% is generally considered a weak oligopoly and a CR_4 of lower than 25% indicating absence of oligopoly

RESULTS AND DISCUSSION

It was observed that most of the milk in the area is marketed in raw form. Results indicated that on average 40% of the total cattle owned by smallholder pastoralist produce milk annually. The cows are 99% of indigenous breed type. The peak production period was the wet season in which about 1.3 litres per cow was produced per day, while about 0.7 litres was produced per cow during the dry season. The annual mean milk produced per day per cow was 1 liter. Furthermore, the results show that the milk reaching the market during the wet season (peak production period) was 27% while during the dry season it was 63% as reflected in Table 2.

Production costs were computed by finding the major costs a producer incured in milk production. They include costs of labour for hearding, dipping/spraying and medication (drugs).The average cost of production per liter was found to be TZS 120(Refer Figure 1) and the average marketing costs for a pastoralist (producer) per

Table 1. Number of mean cattle holding size and mean milk production of sample household pastoralists

Variable Category	Central Zone	Northern Zone	Southern Zone	District Average
Number of Cows	48	20	32	28
Number of indigenous cows	47	20	32	28
Number of dairy Cows	1	0	0	0
Total number of cattle	116	48	83	70
Number of Cow milked in Wet season	23.73	13.10	9.44	14
Number of cow milked in Dry season	23.62	11.83	7.38	12
Quantity of milk produced per day in Wet season(Litres)	35.97	17.91	9.76	18.11
Quantity of milk produced per day in Dry season(Litres)	15.26	9.26	5.44	8.87

Source: Survey results.

Table 2: Mean milk yield per day and market share of pastoral household

Items	Season	Amount (Litres)
Mean milk yield (Liter)	Wet season	18.11
	Dry season	8.87
Mean milk sold (Liter)	Wet season	18.11
	Dry season	5.08
Percentage (%) share of milk marketed	Wet season	27.08

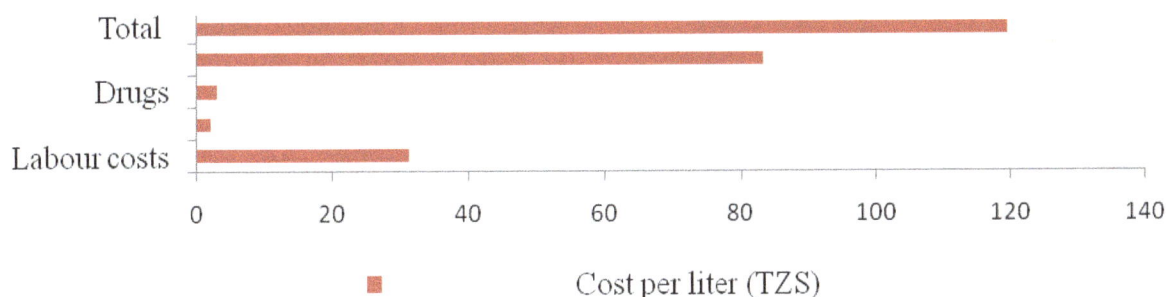

Figure 1. Average production cost/liter of Milk for pastoralists

liter is TZS 66 while the average selling price is TZS 518 as per table 3.The highest production cost pastoralist incur is labour cost which accounts for 26% of the total production costs while transport costs was 51% of the total marketing costs. On average a producer fetched a net profit of TZS 332.

The average marketing cost incurred by a milk vendor for one liter of milk from producers to reach retailer was TZS 61 out of which transport cost was the major component which covered about 59% of the total marketing costs. On an average, a milk vendor got a net profit of TZS 65 as shown in Table 4.For the case of retailer in Table 5, the marketing cost was TZS 193/liter of which labour cost was the highest cost found which covered 58% of the retailer's marketing cost whereas their net profit was found to be TZS 141 per liter. District level collection here

referred as taxi is paid by retailers while the vendors do not pay. The reason might be that retailers have constant buying and selling places and could be easily taxed by the district council regularly where by vendors relatively have no fixed places and are difficult to taxi them regularly.

The study showed that pastoral milk marketing was profitable. Net margins realized were TZS 332, TZS 65 and TZS 141 for the producers, vendors and retailers, respectively. The producer share was 55%, and 45% of the total gross marketing margin was added to milk price when reaches the final consumers as per table 7. The profit of pastoralists per liter of TZS 332 was greater than the profit obtained by vendors and retailers'.Although a pastoralist obtained higher profit than traders, it was very

Table 3. Mean milk marketing cost/liter for pastoralists

Cost item	Cost per liter(TZS)
Transport cost	33.18
Storage costs	4.32
Packaging costs	8.71
Measuring costs	0.11
Spoilage	1.18
Other costs	18.11
Total costs	**65.61**

Table 4. Average cost and Profitability of pastoral milk vendor per liter

Item	Costs(TZS)
Average Buying price	408.82
Average Selling price	535.29
Transport costs	36.06
Storage costs	6.25
Packaging costs	7.81
Labour costs	10.52
Total costs	60.64
Marketing margin	126.47
Net Marketing margin (profit) per liter	**65.83**

Table 5. Average costs and profitability of pastoral milk retailers per liter

Item	Costs (TZS)
Average Buying price	607.69
Average Selling price	942.31
Transport costs	31.09
Storage costs	2.91
Packaging costs	7.81
Measuring costs	0.18
Tax(District collection)	5.41
Spoilage	2.05
Labour costs	112.65
Other costs	31.27
Total costs	193.37
Marketing margin	334.62
Net Marketing margin (profit) per liter	**141.25**

difficult to compare their profits due to the fact that pastoralist obtained this profit for all his/her efforts on livestock husbandry and marketing practices, while small traders and retailers obtained that much of profit in very short time.

This study also employed the Structure, conduct and performance (S-C-P) model from the theory of industrial organization in order to examine the competitiveness of milk market, behavior of the marketing actors and success in meeting their respective goals in the study areas. The pastoral milk market structure of the study area was found to involve marketing agents like pastoralists as producers, small traders/vendors (semi-whole sellers), retailers, milk bars, restaurants, kiosks, milk collection centers for milk market. Consumption of milk is higher in lower age group (8 years and below) than other age groups. The implication of these facts is the growth of milk demand due to increase in lower age population segment being larger consumers of the product.

They preferred fresh milk as it was found during the survey by 76%, 20% and 4% for fresh milk, milk powders and condensed milk respectively. This suggests that even

Table 6. Average milk marketing profit/liter

Milk marketing actor	Marketing cost and profit	Cost (TZS)
Pastoralists	Selling price	518
	Production cost	120
	Marketing cost	66
	Profit	332
Small traders	Purchasing price	409
	Selling price	535
	Marketing cost	61
	Profit	65
Retailers	Purchasing price	608
	Selling price	942
	Marketing cost	193
	Profit	141

Table 7. Average price of milk at different market levels and % share of consumer price

Milk marketing actor	Selling price(TZS)	%(gross marketing margin	Profit per liter (TZS)
Pastoralists	518	55	332
Small traders	535	15	65
Retailers	942	30	141

TGMM (Complete distribution channel) =45%
GMM (Vendors) = 15%
GMM (Retailers) =30%
GMMp (Producer participation) = 55%

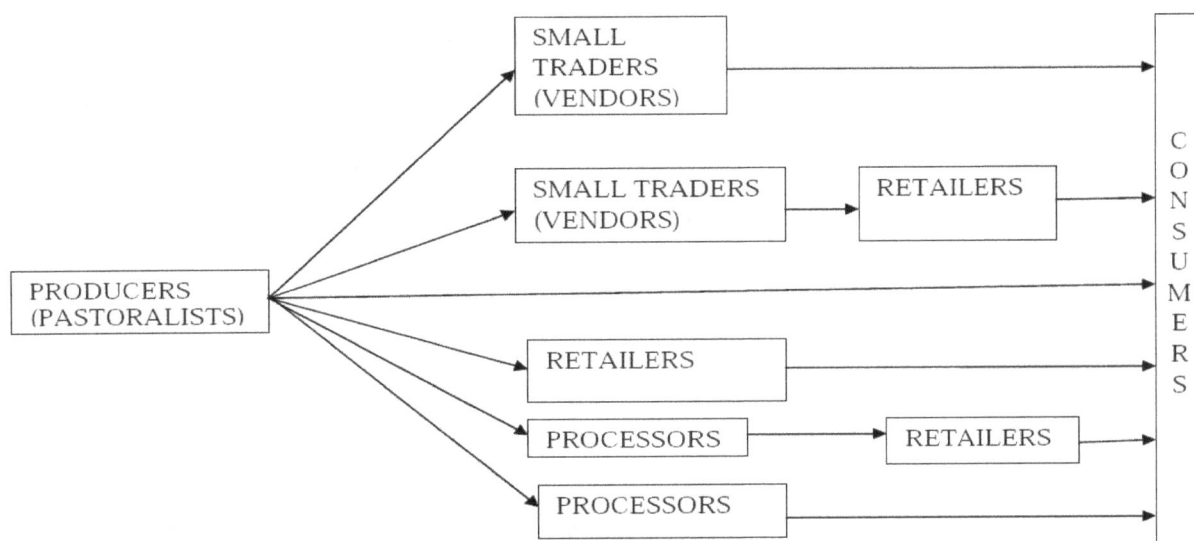

Figure 2. Marketing Channels for Pastoral milk

with importation of fresh milk substitutes, demand for fresh milk was not affected.

The milk market concentration ratio in the study area was recorded at 37.67% suggesting a weak oligopoly market type. Their main source of marketing information among respondents was direct observation of market prices, friends, other vendors through telephone and personal contact, consumers/buyers, or a combination of two or more information sources. Moreover, personal visit to market places and consumers were found to be the most important market information sources of milk during the survey period.

Managerial skills, working capital, nature of commodity and demand and supply conditions, legal and policy constraints were considered in analyzing barriers to milk market entry and exit. The results suggest that 39% of milk traders were illiterate, about 57% and 4% of milk traders had primary and secondary school education respectively. This result portrays that formal education seemed to create no barrier to entry to milk market. Results showed that most of the traders were in milk trading business for more than 6 years. About 25%, 9%, 27%, 32% and 7% of milk traders had the sources of working capital from parents, loan, other businesses, farming and other sources respectively. Since the majority of milk traders had their own source of capital for the trading activities and were able to purchase 2-46 liters of fresh milk per day, lack of capital did not seem to be a constraint for milk market entry. There was no strong restriction to enter into milk marketing with respect to license as 93% small traders and retailers undertaking milk trading without having license. Wholesale markets were relatively free to enter the market as far as they had required amount of capital.

About 47% of the respondents claimed that their major milk marketing problem was seasonality of demand associated with highly perishable nature of milk. Focus group discussion with traders confirmed that some milk traders went out of business activities because they incurred losses and got into debts due to high demand fluctuation associated with perishable nature of milk. This suggests that fluctuating demand coupled with perishability nature of milk created appreciable entry barrier in the milk marketing system.

With regard to quality, it was observed that there were no set standards for milk in the study area. The only quality aspect considered by almost all traders was milk density that gave indication about amount of water in the milk. Another element of quality in the system was odour that traders tested by smelling the milk to assess the extent of milk spoilage. Milk was neither standardized nor graded. Due to lack of standard and grades, buyers determined price of milk through visual inspection. Furthermore, there was no standard measuring and packaging tool for milk. Materials found were plastic buckets and tins usually belonging to traders, calabashes containers and pots on the part of producers.

Most of the pastoralists sold their produce at farm level. Those who did not sell their produce at farmer level for different reasons transported milk as headload or used bicycle to move milk to nearest market. It was observed that 82%, 2% and 16% of producers transported milk using bicycle, buses and others (motorcycles and cars) respectively. Storage facilities were poor at all levels due to lack of electric power. In this respect, only 25% of milk vendors stored milk in refrigerators while the remaining 75% lack the cold storage facilities to increase the shelf life of the milk.

Pastoralists retain part of the milk for home consumption and sell the rest in the market to get cash income. Milk production per household was found to be related to the number of cows. During the dry season households sold about 63% of total milk produced per day, which was higher than average of the wet season 27%. Focus group discussions with stakeholders highlighted that producers sold milk to their regular customers although they had no permanent relationship or contractual arrangements between sellers and buyers. Relationship existed at the time of transaction only. Nonetheless in some cases verbal agreement existed, but these had no legal powers. The milk marketing system was in most cases characterized by no licensing requirement to generate the operation, low cost of operation and no regulation of operation. Results indicated that the bulk of milk was marketed through traditional channels and transactions found to take place with direct contact between seller and buyer. There existed no broker in the milk marketing system. With regard to contractual agreement between market actors, verbal agreement seemed to prevail for quality and supply assurance of pastoral milk.

It was revealed that factors such as seasonality (dry and wet) exerted considerable impact on milk price determination. The price of milk reached peak during dry seasons and dropped substantially during lean (wet) season. The pricing structure of milk had little to do with milk quality. Price level was observed to be determined by forces of demand and supply. Increase in supply was mainly linked to water supply whereas the quantity of milk demanded depends on factors influencing milk consumption in big cities such as Tanga, Dar es Salaam and Morogoro town.

Market information supply was not transparent between levels that created price discrepancy and differences among actors. Milk vendors had better price information access from their friends while other actors did not have the access leading to information irregularity. Product selling and potential buyers accessing their areas were the main market information producers' used. Producers were taking the market price while the milk vendors were the dominant source of information. The producers were price takers without negotiation. The lack of modernized post harvest handling practices and short shelf life of milk forced producers to sell at their prevailing prices. Knowing this, small traders exerted pressure to producers to sell at low price. Addition of water by small traders and some of the producers was one of unethical practice followed by different marketing actors. Marketing costs and margins were used to determine whether there were excess profits and serious inefficiencies or whether wide margins are due to technical constraints. The profit of pastoralists per liter suggests that there was a profit of TZS 332/liter which was greater than the profit obtained by small traders and retailers'. This situation implied that there was good performance.

Break even analysis was used to find the break even output per animal. The results showed that for a pastoralist to break even needs to produce an output of 1140 liter/cow .According to Njombe *et al,* (2011) indigenous cattle produced around 300 - 500 liters/lactation which show that in order to break even a cow should lactate 2-4 times. By considering also the gestation period a cow should be kept for milk production for three to five years in order for a pastoralist to recover all the costs.

CONCLUSION

This paper intended to analyze and document the efficiency of the pastoral milk marketing system so as to recommend the strategies for sufficient marketing with the view of the market structure, conduct and performance (profitability) of the system in Kilosa district. Results from this study indicate that pastoral milk marketing was not efficient but profitable and not competitive. The profit margins were found to vary vertically across the chain. Pastoralists received higher returns than small traders and retailers. The market information flow was non-transparent and not helpful to all market participants. It was also observed that the system had no defined standards and grades and no product differentiation in terms of defined packaging by both pastoralists and traders signaling market inefficiency. Pastoralists and traders have no organizations where they can collectively influence prices.

Pastoralists' small holders were found to own cattle with the average of 70 per household. Marketing could be a means of income providing business opportunities for all actors in the value chain. This implied that the pastoralists are in better position to change the economy of Kilosa district if the pastoral milk market will be organized and operate efficiently. The government attention is needed in improving the inefficient marketing system. Commercialization focusing on promotion of pastoral milk marketing, through introduction of community based extension services and village community banks, empowerment of small and medium scale processors, training on proper milking and milk handling; as well as group formation and organizations all need to be considered in the future intervention strategies.

ACKNOWLEDGEMENT

We would like to extend much gratitude to project entitled 'Milk and Beef Value Chains in the Pastoral Livestock System, Kilosa District" under Enhancing Pro-poor Innovation in Natural Resources and Agriculture Value Chains (EPINAV) program which funded this research work. We are grateful to project team leader Prof. Esron D. Karimuribo and co-project team leader Prof. Emmanuel K. Batamuzi and all the members of the project team for their unconditional support and contributions. The authors acknowledged the contributions of Dr. Sylvia Angubua Baluka, ASHISH PARASHARI, Dr. Mohammed Salem, Pramod Kumar Mishra ALMAZ GIZIEW, Dr. Kheiry Hassan M. Ishag, Dr. Prakashkumar Rathod, Dr. Workneh Abebe Wodajo, Chisoni Mumba and Ritesh Patel for donating their time, critical evaluation, constructive comments, and invaluable assistance toward the improvement of this very manuscript.

REFERENCES

FAO (2009). *State of Food and Agriculture. Livestock in the Balance.* Food and Agriculture Organization, Rome, Italy. 168pp.

Kohls RL, Uhl JN (1985). *Marketing of Agricultural Product.* McMillan Publishing Company, New York, USA. 52pp.

Mari FM. (2009). Structure and efficiency analysis of vegetable production and marketing in Sindh, Pakistan. Thesis for Award of PhD Degree at University of Tando Jam, Sindh, Pakistan, 223pp.

Mdoe N, Mnenwa R (2007). Study on Options for pastoralists to secure their livelihoods. Assessing the total economic value of pastoralism in Tanzania. A report submitted to TNRF.

Mung'ong'o C, Mwamfupe D. (2003). *Poverty and Changing Livelihoods of Migrant Maasai Pastoralists in Morogoro and Kilosa Districts, Tanzania.* Mkuki na Nyota Publishers, Dar es Salaam, Tanzania. 41pp.

Njombe AP, Msanga YN. (2008). *Livestock and Dairy Industry Development in Tanzania.* Department of Livestock production and Marketing Infrastructure Development. Ministry of Livestock Development and Fisheries, Dar es Salaam, Tanzania. 16pp.

Njombe AP, Msanga Y, Mbwambo N, Makembe N. (2011). The Tanzania dairy industry: Status, opportunities and prospects. *Paper Presented to the 7th African Dairy Conference and Exhibition held at Moven Pick Palm Hotel,* Dar es Salaam, Tanzania. 25 – 27 May, 2011. 19pp.

Sheik S, Edwards S, Allen AJ (2005).Market Structure Conduct Performance (SCP) Hypothesis Revisited using Stochastic Frontier Efficiency Analysis. *Selected paper prepared for presentation at the American Agricultural Economics Association Annual meeting*, Long Beach, California, July 23-26.2006, 21pp.

Tenga R, Mattee A, Mdoe N, Mnenwa R, Mvungi S,Walsh MT (2008). A Study on Options for Pastoralists to Secure their Livelihoods in Tanzania: Current Policy, Legal and Economic Issues. Volume One: Main Report. Report commissioned by CORDS, PWC, IIED, MMM Ngaramtomi Centre, TNRF and UCRT, and funded by CORDAID, TROCAIRE, Ireland Aid and the Wildlife Conservation Society AHEAD Programme.

United Republic of Tanzania (2010). *Economic Survey.* Ministry of Finance, Dar es Salaam, Tanzania. 271pp.

United Republic of Tanzania (2005). *Livestock Sample Survey Census.* National Bureau of Statistics, Dar es Salaam, Tanzania. 75pp.

World Bank (2008). Agriculture for development. The World Bank, Washington DC.

Scoring marketing obstacles and recommended solutions: the case of the date marketing system in Saudi Arabia

Sobhy M. Ismaiel[1], Ahmed M. Al-Abdulkader[2], Safar H. Al-Kahtani[1], Ali I. Saad[1]

[1]Department of Agriculture Economics, College of Agricultural Sciences, King Saud University, Riyadh-Saudi Arabia.
[2]National Plan for Science, Technology and Innovation, King Abdulaziz City for Science and Technology, Riyadh- Saudi Arabia.

This study is an attempt to identify problematic issues facing date marketing systems in Saudi Arabia and to introduce corrective solutions to overcome them. Eventually, this would increase date marketing efficiency, performance, and the competitiveness level of Saudi dates, both locally and globally. The study applied a typical five-level Likert scale and factor analysis in order to analyze the primary data collected from date dealers at the major Saudi Arabian date markets. Results highlighted that the impact of technical problematic issues were the highest, with an average of 3.67 followed by structural problematic issues with an average of 3.36, and finally followed by behavioral problematic issues with an average of 3.12. Technical, structural, and behavioral problematic issues have been categorized into a set of factors explaining 79.18%, 74.84%, and 77.7% of the variations, respectively. The impact of applying the recommended solutions on a Saudi Arabian date marketing system were found to be high, ranging from a minimum of 3.35 for encouraging competition to a maximum of 3.98 for providing affordable cold storages. Dates marketing channel types, geographical factors, and the date marketing dealers' educational level have shown significant effects on the prevailing problematic issues of the Saudi Arabian date marketing system.

Keywords: Marketing efficiency, Structural problems, Behavioral problems, Technical problems, Likert scale, Factor analysis.

INTRODUCTION

The date sectors an auspicious sector in Saudi Arabia, with a high trading volume that is equal to about US$ 6 billion annually. Improving date marketing efficiency and confronting its obstacles and problems is likely to increase the annual trading volume of the Saudi dates by about 30 %(Al-Abdulkader et al., 2016). The marketing problems and obstacles are tremendous, and their prevalence and persistence hinder the marketing system advancement and the improvement of marketing efficiency. The chronic problems of the agricultural marketing in developing countries include inadequate storage and transportation facilities, poor market information system, and capital shortage. Such problems require integrated national measures that serve producers, consumers, and national development goals, including, a high definition of technical assistance and research needs, adequate capability of marketing management, and wide spectrum marketing information to all beneficiaries (Friedberg, 1970).The problems and obstacles of agricultural marketing are classified according to their relation to market structure, conduct, and performance.

*Corresponding Author: Ahmed M. Al-Abdulkader, National Plan for Science, Technology and Innovation, King Abdulaziz City for Science and Technology, Riyadh-Saudi Arabia. akader@kacst.edu.sa

They could be caused by the technical issues of the provided marketing services (Ismail and Alshehry, 2015). Date marketing dealers, including, wholesalers, retailers, and exporters are capable of identifying the problems facing their marketing business and negatively affecting their marketing efficiency, as well as providing corrective solutions according to their own attitudes, understanding, and experience. Many variables affect marketing efficiency either in short or long term, such as return on investment, internal rate of return, economic value, competition over time, profit, and others (Rust et al., 2004, Sheth et al., 2002). Major marketing constraints are diversified in terms of lack of reliable market information systems, low technological know-how for value chain upgrading and development, low bargaining power of farmers (Emana et al., 2015)lack of competitiveness at the upstream marketing bodies, and lack of technological content (Han, 2011). Governance, politics, and infrastructure are challenges in the agricultural marketing (Voboril, 2013).

Agricultural marketing systems camber addressed from different perspectives, including producers, marketing efficiencies, prices, margins, integration, and equilibrium, in both place and time dimensions. It can be further addressed from the perspective of marketing in its form dimension, marketing risks and information. Another viable perspective includes the methods used to improve agricultural marketing systems (Alkahtani et al., 2007), as well as the basic elements of the agricultural marketing system itself, such as marketing margins, costs, and profits for fruits and vegetables commodities (Alkahtani and Elfeel, 2006; Hassan and Raha, 2013). Still another would be there commendations given to improve date marketing systems, such as by increasing the producers' share of date retail prices in order to recognize the basic factors responsible for the high marketing margins of some vegetables and fruits on the retailing level (Ismail, et al., 2009), in addition to the application of e-trading in date marketing for the sake of identifying problematic issues that hinder the application of electronic trade systems and, thus, increase marketing efficiency and reduce the costs for providing products and contribute to an increase of date marketing volume (Alkahtani et al., 2011). Date marketing dealers' level of satisfaction with the provided Saudi Arabian marketing service costs was assessed against an average, ranging between a minimum of 2.62 for non-cold storage and a maximum of 2.71 for packaging, grading, and sorting, out of 5.00 on the Likert scale, with the exception of transportation, which was satisfactory with an average of 2.53. On the other hand, the date marketing dealers 'satisfaction with the provided marketing service quality was assessed at an average ranging between 2.61 and 3.40 out of 5.00 on the Likert scale, except for transportation costs, which were satisfactory with an average of 2.59 (Al-Abdulkader et al., 2016).

This study is abstracted from an original comprehensive research project that dealt with the marketing of dates in Saudi Arabia from different dimensions (Al-abdulkader et al., 2015), and its main purposes are to identify the problems and obstacles that the Saudi Arabian date marketing system faces and to introduce recommended solutions to overcome these issues, thus improving date marketing efficiency and performance, as well as enhancing Saudi Arabian date competitiveness, both locally and globally.

MATERIALS AND METHODS

A field survey questionnaire was conducted on a random sample of date marketing dealers functioning at the major Saudi Arabian date markets, namely, Riyadh, Al-Madinah Al-Munawara, Al-Hasa, and Al-Qassim. This random sample covered 298 date marketing dealers representing all types, namely, wholesaling, retailing, and exporting. Table 1 presents the study sample's main characteristics. Qualitative and quantitative statistical analyses were applied in order to analyze the collected primary data. The qualitative statistical analysis measures the means, standard deviations, frequency, and variable frequency in order to identify the functional and personal characteristics of study sample and their questionnaire feedback. In addition, a typical five-level Likert scale was applied in order to determine the average impact record of problematic issues on the Saudi Arabian date marketing system, as follows:

No Impact: shows a very low degree of impact on the date marketing system from the respondents' view, with grades ranging between1 and 1.80.

Impact to some extent: shows a low degree of impact on the date marketing system from the respondents' view, with grades ranging between 1.81 and 2.60.

Moderate impact: shows a certain degree of impact on the date marketing system from the respondents' view, with grades ranging between 2.61 and 3.40.

High impact: shows a high degree of impact on the date marketing system from the respondents' view, with grades ranging between 3.41 and 4.20.

Very high impact: shows a very high degree of impact on the date marketing system from the respondents' view, with grades ranging between 4.21 and 5.00.

Factor analysis was also applied in order to reduce the large number of date marketing problems and solutions into a smaller set of unobserved categories called factors, with minimum information loss. The principle component and various rotation methods, which are most commonly used in factor analysis, were applied in order to identify the most important factors explaining correlation patterns within the set of observed variables. Factor analysis seeks to identify common underlying factors out of observed variables, so that: 1) the first factor is the one most correlated with the observed variables or is the

Table 1. Basic Characteristics of the Study Sampleat Major Saudi Arabian Date Market Regions

Variable	Frequency	%
Study Regions		
Riyadh	75	25.2
Al-Madinah Al-Munawara	101	33.9
Al-Hasa	49	16.4
Al-Qassim	73	24.5
Types of Marketing Channels		
Wholesaling	47	15.8
Retailing	129	43.3
Wholesaling and Retailing	106	35.6
Retailing and Exporting	3	1.0
Wholesaling, Retailing, and Exporting	6	2.0
Unidentified	7	2.3
Specialization in Date Marketing Activities		
Specialized	216	72.5
Non-Specialized	82	27.5
Other Marketing Activities Rather Than Dates (for Non-specialized)		
Agricultural Marketing Activity	71	86.59
Non-Agricultural Marketing Activity	11	13.41
Date Marketing History of the Dealer		
Inherited business	116	38.6
Non-inherited business	182	61.1
Educational Level ofDate Marketing Manager		
Lower than Secondary education	150	50.3
Secondary Education	120	40.3
Higher Education	24	8.1
Unidentified	4	1.3
Scale of Trade (ton)		
Small scale < 10	51	17.2
Medium scale 10 –< 50	125	41.9
Above medium scale 50 – < 100	33	11.0
Large scale 100 – 500	67	22.5
X-large scale > 500	22	7.4

Source: Study Sample, 2012

most explained of the common variation, followed by the second factor and so on; 2)in every factor there are a few non-zero coefficients; and 3)these factors can be interpreted in line with their observed variable relation. Communality is an observed variable correlation with more than one factor. The degree of variable communality is the variable contributions of all factors, and it measures the percent of variance in the observed variable explained by all factors. Factor loading consists of the correlation between observed variables and factors. Squared factor loadings denote the percentage of the observed variable variance that is interpreted by the factor. Eigen values are the variance of the factor. The first factor accounts for the highest variance, followed by the second factor and so on. If the Eigen value is greater than one, the factor will be accepted; otherwise, it will be refused as a factor. Factor rotation is a process of mathematical procedures performed to achieve a simple and practically meaningful factor solution.

RESULTS AND DISCUSSIONS

Problems and Obstacles

The key issues in the Saudi Arabian date marketing system have been categorized into three dimensions: technical, structural, and behavioral. Technical issues have the highest impact on the date marketing system, with an average of 3.67 on a Likert scale and a standard deviation of 0.84, followed by structural issues, with an average of 3.36 and a standard deviation of 0.74. Behavioral issues average 3.12, with a standard deviation of 1.03. The general average for all dimensions 3.38.

Technical Issues

The impact of technical issues on the date marketing system were found to be high, ranging from a minimum

Table 2. Key Technical Issues in the Saudi Arabian Date Marketing System from the Date Marketing Dealers' Perspective

Problematic Issue		Very high impact	High impact	Moderate impact	Impact to some extent	No impact	AVG	STDEV	Rank
1. Lack of adequate grading and sorting services	F	54	119	67	45	12	3.53	1.08	7
	%	18.1	39.9	22.5	15.1	4			
2. Lack of adequate packaging services	F	63	121	68	38	8	3.65	1.03	5
	%	21.1	40.6	22.8	12.8	2.7			
3. Lack of adequate transportation and distribution services	F	59	128	71	31	8	3.67	1.00	4
	%	19.8	43	23.8	10.4	2.7			
4. Lack of adequate storage services	F	77	121	62	26	6	3.81	0.99	1
	%	25.8	40.6	20.8	8.7	2			
5. Slow procedures for exporting	F	74	85	94	34	8	3.62	1.07	6
	%	24.8	28.5	31.5	11.4	2.7			
6. High costs of marketing services	F	90	94	68	33	12	3.73	1.13	2
	%	30.2	31.5	22.8	11.1	4			
7. Low-qualitymarketing services	F	82	91	77	32	9	3.70	1.09	3
	%	27.5	30.5	25.8	10.7	3			
Dimension Average							3.67	0.84	

Source: Study Sample, 2012

impact of 3.53 for the lack of adequate grading and sorting services to a maximum of 3.81 for the lack of adequate storage services (see Table 2).The prevailing technical issues of date marketing system in Saudi Arabia, mainly, the lack of adequate transportation and distribution services, and storage services were considered among the chronic marketing problems prevailed in developing countries (Kries berg, 1970).

Structural Issues

The impact of structural issues were found to be high to moderate, ranging from a minimum of 3.8 for the monopoly in marketing services to a maximum of 4.01 for the shortage of seasonal laborers and the occurrence of black laborers market where the scarce seasonal laborers are overpriced (see Table 3). The impacts of observed structural issues are categorized as follows:
Structural issues with a high impact (3.41–4.20):
Deficiency of season all a borers and the emergence of a black laborer market.
The emergence of non-local purchasing power financing date marketing operations.

Increased supply.
Lack of clear quality standards and specifications.
Insufficient packaging and process in plants.
Lower demand.
Structural issues with a moderate impact (2.61–3.40):
Lack of adequate marketing information.
Barriers to market entry and exit.
Existence of a monopoly (in quantity and price).
Monopoly of marketing services.
Behavioral Issues:
Behavioral issues were found to have a moderate impact on the Saudi Arabian date marketing system, with an a minimum of 3.05 on a Likert scale for poor knowledge of price determination and a maximum of 3.22 for lacking knowledge on quality discrimination (see Table 4).

Factor Analysis for Studied Issues
Factor Analysis for Technical Issues
The First Factor

The Eigen value for this factor amounted to 4.5, and it explained 43.77% of the variations in technical date marketing problems. This ratio was higher than the other

Table 3. Key Structural Issues in the Saudi Arabian Date Marketing System from the Date Marketing Dealers' Perspective

Problematic Issue		Very high impact	High impact	Moderate impact	Impact to some extent	No impact	AVG	STDEV	Rank
1. Monopoly in price/quantity	F	37	66	65	69	60	2.84	1.32	9
	%	12.4	22.1	21.8	23.2	20.1			
2. Monopoly in marketing services	F	26	66	81	68	55	2.80	1.23	10
	%	8.7	22.1	27.2	22.8	18.5			
3. Lack of adequate information system	F	27	77	83	69	42	2.93	1.19	7
	%	9.1	25.8	27.9	23.2	14.1			
4. Barriers to market entry and exit	F	30	73	76	69	48	2.89	1.24	8
	%	10.1	24.5	25.5	23.2	16.1			
5. Low demand	F	62	111	70	35	19	3.55	1.14	6
	%	20.8	37.2	23.5	11.7	6.4			
6. Increase supply	F	73	112	59	34	20	3.62	1.17	3
	%	24.5	37.6	19.8	11.4	6.7			
7. Inadequate dates packaging and processing factories	F	73	105	57	38	22	3.58	1.22	5
	%	24.5	35.2	19.1	12.8	7.4			
8. The absence of clear quality specifications and standards	F	79	95	61	46	15	3.60	1.18	4
	%	26.5	31.9	20.5	15.4	5			
9. Shortage of seasonal laborers and the emergence of a black laborer market	F	126	95	41	27	9	4.01	1.09	1
	%	42.3	31.9	13.8	9.1	3			
10. The emergence of non-local purchasing power financing dates marketing operations	F	102	87	43	36	17	3.78	1.23	2
	%	34.2	29.2	14.4	12.1	5.7			
Dimension Average							3.36	0,74	

Source: Study Sample, 2012

factors. This factor includes four variables, namely, shortage of grading, shortage of packaging, shortage of transportation, and shortage of storage. The correlation coefficient of these items was to the factor of 0.808, 0.840, 0.847, and 0.684, respectively. The degree of commonality ranged between 0.710 for shortage of storage and 0.878 for shortage of packaging. These results confirm that all of the variables aforementioned

reach a satisfactory level from the date dealers' perspective.

The Second Factor

The Eigen value for this factor amounted to 1.03, and it explained 35.96% of the variations in technical date marketing problems. This factor includes three variables,

Table 4. Key Behavioral Issues in the Saudi Arabian Date Marketing System from the Date Marketing Dealers' Perspective

Problematic Issue		High impact	impact	Moderate impact	Impact to some extent	No impact	AVG	STDEV	Rank
1. Poor knowledge of price determination	F	29	86	90	56	37	3.05	1.17	4
	⁒	9.7	28.9	30.2	18.8	12.4			
2. Poor knowledge of best markets	F	26	86	108	44	34	3.09	1.11	3
	⁒	8.7	28.9	36.2	14.8	11.4			
3. Lack of knowledge in marketing ways	F	37	92	84	50	35	3.15	1.19	2
	⁒	12.4	30.9	28.2	16.8	11.7			
4. Lack of knowledge on quality discrimination	F	48	83	84	47	34	3.22	1.23	1
	⁒	16.1	27.9	28.2	15.8	11.4			
Dimension Average							3.12	1.03	

Source: Study Sample, 2012

Table 5. Factor Analysis for the Elements of Saudi Arabian Technical Marketing Problems

Elements	Factors		Communality
	First factor	Second factor	
1. Lack of adequate sorting and grading	0.808		0.869
2. Lack of adequate packaging	0.809		0.878
3. Lack of adequate transportation	0.847		0.866
4. Lack of adequate storage	0.684		0.710
5. Slow procedures for exporting		0.641	0.763
6. High cost of marketing services		0.867	0.880
7. Low-quality marketing services		0.855	0.873
Eigenvalue	4.50	1.03	-
Explained variation by elements (percent)	43.22	35.96	79.18
Cronbach's alpha	0.904	0.87	-

Source: Study Sample, 2012

namely, complications of exporting techniques, high cost of marketing services, and low quality of marketing services. The correlation coefficient of these items was to the factor of 0.641, 0.867, and 0.855, respectively. The communality factor ranged from0.763 for complications of exporting techniques to0.880 for high cost of marketing services. These results confirm that all of the aforementioned variables have the same level from the date dealers' viewpoint. Logistic regression shows no effect of any explanatory factors on technical marketing problems (see Table 5).

Factor Analysis for Structural Issues
The First Factor

The Eigen value for this factor amounted to 4.00 and explained 29.71% of the marketing problem variations. This ratio was the highest compared to the other factors. This factor includes a monopoly on quantity and price, a monopoly on date marketing services, a lack of information, and a limited market entrance or exit. The correlation coefficient of these items to the factor was of 0.805, 0.822, 0.702, and 0.647, respectively. The

Table 6. Factor Analysis for the Elements of Structural Problems inthe Saudi Arabian Date Marketing System

Elements	Factors			Communality
	First factor	Second factor	Third factor	
1. Monopolistic power (quantity and price)	0.805			0.888
2. Monopolistic power (marketing services)	0.822			0.901
3. Insufficient marketing information	0.701			0.814
4. Limitations of market entrance	0.647			0.794
5. Low demand		0.760		0.841
6. Excess supply		0.829		0.894
7. Insufficient data-processing facilities		0.656		0.716
8. Unclear quality standards		0.669		0.682
9. Deficiency of seasonal laborers			0.766	0.851
10. Foreign financial capacity for financing marketing services			0.828	0.874
Eigenvalue	4.0	2.55	0.926	
Explained variation by elements (percent)	29.71	27.01	18.1	74.84
Cronbach's alpha	0.88	0.86	0.76	-

Source: Study Sample, 2012

communality factor ranged between 0.794 for limitations to entry or exit and 0.901 for monopoly in marketing services. These results confirm that all of the aforementioned variables have the same scoring level from the date dealer viewpoints.

The Second Factor

The Eigen value for this factor amounted to 2.55 and explained 27.01% of the date marketing problem variations. This factor includes demand decline, excess supply, insufficient manufacturing plants, and inadequate quality standards. The correlation coefficient of these items was to the factor of 0.760, 0.829, 0.656, and 0.669, respectively. The communality factor ranged from 0.682 for inadequate quality standards to 0.894 for excess date supply. These results confirm that all of the aforementioned variables have the same satisfactory level from the date dealer viewpoint.

The Third Factor

The Eigen value for this factor amounted to 0.926, and it explained 18.1% of the variations in date marketing

service satisfaction. This factor includes shortage of seasonal labor and non-Saudi financial sources for date marketing services. The correlation coefficient of these items was to the factor of 0.766 and 0.828, respectively. The communality factor ranged between 0.815 and 0.874, respectively. These results confirm that all of the aforementioned variables have the same level from the date dealers' viewpoint. Logistic regression shows that the type of date merchant activity (type of date marketing channel) only affects the score given for structural problems from the date dealers' perspective (see Table 6).

Factor Analysis for Behavioral Issues

The Only Factor: Behavioral Problems

The Eigen value for this factor amounted to 3.12, explaining 77.7% of the variations in date marketing service satisfaction level. This factor includes all four variables in one factor. The correlation coefficient of these items relates to the factor was0.853, 0.919, 0.914, and 0.837, respectively. The Cranach alpha factor for stability is 0.902. This factor represents behavioral problems.

Table 7. Recommended Solutions to Improve Date Marketing Efficiency from the Perspectives of Date Marketing Dealers in Saudi Arabia

Recommended Solutions	AVG	STDEV	Rank
Providing affordable cold storages	3.98	1.00	1
Providing marketing loans	3.92	1.06	2
Monitoring auctions	3.84	1.06	3
Stimulating the role of cooperative societies and strengthening the governmental support	3.79	1.08	4
Founding investment portfolios for enterprisers' youth training and financing activities of date purchasing and marketing	3.67	1.13	5
Escalating institutional arrangements for date marketing locally and globally	3.74	1.00	6
Establishing a national federation for dates producers consisting of private sector, public sector, universities, and research centers	3.74	1.04	7
Establishing dates' marketing companies	3.74	1.15	8
Expanding dates' complementary industries	3.71	1.03	9
Subsidizing exportation	3.65	1.04	10
Enforcingof specifications and standards	3.64	1.00	11
Regulating wholesale markets	3.57	1.05	13
Ensuring the availability of market information	3.57	1.07	13
Promoting date marketing techniques, such as electronic dates and auctions	3.57	1.10	14
Date manufacturing	3.55	1.02	15
Intensifying awareness and counseling programs	3.50	1.02	16
Encouraging competition	3.35	1.15	17
Dimension Average	**3.68**	**0.71**	

Source: Study Sample, 2012

Recommended Solutions

The impact of the recommended solutions for overcoming or minimizing the impacts of the Saudi Arabian date marketing system issues, thus enhancing the marketing efficiency, are almost high. They ranged between a minimum of 3.35 for encouraging competition and a maximum of 3.98 for providing affordable cold storages, with an average of 3.68 (see Table 7). Al-abdulkader et al. (2015, 2016) found that solving and minimizing the impacts of these issues would enhance the date marketing efficiency by about 51%.
Saudi Arabia

Factor Analysis for Solutions

Factor analysis suggested categorizing the set of

recommended solutions for date marketing problems into four factors to improve date marketing efficiency. These four factors explain 71.02% of the variations in all of the items of recommended solutions. These four factors are presented in Table (8) and discussed below.

The First Factor

The Eigen value for this factor amounted to 7.67, and it explained 19.8% of the variations in the scores given to the solutions. This ratio was higher than the other factors. This factor includes enhancing the competition, organizing wholesale markets date manufacturing and processing, encouraging date exporting, and improving standardization. The correlation coefficient of these items was to the factor of 0.651, 0.774, 0.663, 0.784, and 0.622, respectively. The communality factor ranged from

Table 8. Factor Analysis for the Elements of Recommended Solutions for Saudi Arabian Date Marketing Problems

Elements	Factors				Communality
	First factor: cost of physical services	Second factor: cost of exchange services	Third factor: cost of facilitating services	Fourth factor: other costs	
1. Enhancing competition	0.651				0.776
2. Organizing wholesale markets	0.774				0.852
3. Date manufacturing	0.784				0.829
4. Supporting date exports	0.663				0.776
5. Encouraging specifications and standards	0.622				0.554
6. Supporting cooperatives		0.700			0.581
7. Establishing national date producers union		0.651			0.644
8. Enhancing date electronic trade		0.736			0.819
9. Complementary manufacturing		0.697			0.777
10. Enhancing institutional framework on both national and international levels		0.677			0.770
11. Low-cost cold storing			0.706		0.744
12. Providing marketing loans			0.772		0.813
13. Establishing date marketing companies			0.662		0.768
14. Training and financing			0.769		0.727
15. Marketing extension				0.740	0.685
16. Marketing information				0.787	0.772
17. Controlling date auctions				0.682	0.728
Eigen value	7.67	2.03	1.35	1.02	
Explained variation by elements (%)	19.8	19.02	18.38	13.82	71.02

Source: Study Sample, 2012

0.554 for improving the date standardizationto0.852 for wholesale market organization. These results confirm that all of the aforementioned variables have the same level of satisfaction from the date dealers' viewpoint.

The Second Factor

The Eigen value for this factor amounted to 2.03, and it explained 19.02% of the variations in date marketing service satisfaction. This factor includes reactivating the role of cooperatives in date marketing; establishing a date producers union; enhancing information technology such as electronic trade in date marketing; expanding date manufacturing and its complementary industries; and improving the institutional framework in the date marketing system, on both local and international levels. The correlation coefficients of these items are to the factor of 0.700, 0.651, 0.697, 0.736, and 0.677,

Table 9. One-Way Analysis of Variance Showing the Impact of Marketing Channel Types on Major Saudi Arabian Date Marketing Problems

Comparison variable	Marketing channels	Average Score	STDEV	F-test value	Significant level	Interpretation
Structural issues	Wholesaling	3.15	0.88	1.79	0.148	-
	Retailing	3.41	0.69			
	Wholesaling, retailing	3.42	0.73			
	Wholesaling, retailing, exporting	3.23	0.84			
Behavioral issues	Wholesaling	2.94	1.04	2.20	0.088	-
	Retailing	3.26	0.97			
	Wholesaling, retailing	3.01	1.06			
	Wholesaling, retailing, exporting	3.58	1.35			
Technical issues	Wholesaling	3.35[B]	0.72	3.26	0.022*	All marketing channels versus wholesaling
	Retailing	3.77[A]	0.81			
	Wholesaling, retailing	3.73[A]	0.83			
	Wholesaling, retailing, exporting	3.77[A]	1.20			
Date marketing general problems	Wholesaling	3.14[B]	0.70	2.84	0.038*	All marketing channels versus wholesaling
	Retailing	3.48[A]	0.65			
	Wholesaling, retailing	3.39[A]	0.67			
	Wholesaling, retailing, exporting	3.53[A]	1.09			
Recommendations for improved date marketing efficiency	Wholesaling	3.43[B]	0.86	3.41	0.018*	All marketing channels versus wholesaling
	Retailing	3.78[A]	0.59			
	Wholesaling, retailing	3.67[A]	0.72			
	Wholesaling, retailing, exporting	3.98[A]	0.76			

*Significant at 0.05 significance level
- Means followed by the same letter do not significantly differ using Duncan's test.
Source: Study Sample, 2012

respectively. The communality factor ranged from0.581 for reactivating the role of cooperatives in date marketing to0.000 for enhancing information technology in date marketing, such as electronic trade. These results confirm that all of the aforementioned variables have the same level of satisfaction from the date dealers' viewpoint.

The Third Factor

The Eigen value for this factor amounted to 1.35, and it explained 18.38% of the variations in the recommended solution scores. This factor includes establishing low-cost cold storage facilities, enabling marketing credit, and establishing dates, marketing companies, and marketing

Table 10. One-Way Analysis of Variance Showing the Impact of Regions under Investigation on Main Saudi Arabian Date Marketing Problems

Comparison variable	Study areas	Average score	STDEV	F-test value	Significant level	Interpretation
Structural issues	Riyadh	3.24	0.87	1.56	0.199	-
	Al-Madinah Al-Munawara	3.36	0.67			
	Al-Hasa	3.54	0.75			
	Al-Qassim	3.34	0.70			
Behavioral issues	Riyadh	2.87^B	1.14	3.55	0.015*	Al-Madinah Al-Munawara versus Riyadh
	Al-Madinah Al-Munawara	3.36^A	0.95			
	Al-Hasa	3.14^{AB}	1.06			
	Al-Qassim	3.04^{AB}	0.97			
Technical issues	Riyadh	3.38^B	0.92	4.75	0.003**	All regions versus Riyadh
	Al-Madinah Al-Munawara	3.80^A	0.72			
	Al-Hasa	3.83^A	0.89			
	Al-Qassim	3.69^A	0.80			
Date marketing general problems	Riyadh	3.16^B	0.79	4.33	0.005**	All regions versus Riyadh
	Al-Madinah Al-Munawara	3.51^A	0.61			
	Al-Hasa	3.50^A	0.70			
	Al-Qassim	3.36^A	0.63			
Recommendations for improved date marketing efficiency	Riyadh	3.67	0.73	0.82	0.486	-
	Al-Madinah Al-Munawara	3.64	0.63			
	Al-Hasa	3.82	0.80			
	Al-Qassim	3.65	0.71			

*Significant at 0.05 significance level
**Significant at 0.01 significance level
- Means followed with the same letter do not significantly differ using Duncan's test.
Source: Study Sample, 2012

training. The correlation coefficients for these items were to the factors of 0.706 0.772, 0.769, and 0.662, respectively. The communality ranged from0.727 for marketing training to0.813 for enabling marketing loans. These results confirm that all of the aforementioned variables have the same level of satisfaction from the date dealers' viewpoint.

The Fourth Factor

The Eigen value for this factor amounted to 1.02, and it explained 13.82% of the recommended solution variations. This factor includes marketing extension, enhancing marketing information, and auction control. The correlation coefficients for these items were to the factor of 0.740, 0.787, and 0.682, respectively. The

communality factor ranged from 0.685 for marketing extension to 0.772 for marketing information. These results confirm that all of the aforementioned variables have the same level of satisfaction from the date dealers' viewpoint.

Impact of Explanatory Factors

The study survey includes some explanatory variables, including marketing channel types (wholesaling, retailing, exporting, and combinations of some or all), regions or geographic factor, date dealers' educational level, the extent of the dealer concentration in date marketing, types of other dealer activity besides date merchandising, and the length of date merchandising experience. One-way analysis of variance and a Chafee test, t-test, and

Table 11. One-Way Analysis of Variance Showing the Impact of Dealer Educational Level on Saudi Arabian Date Marketing Problems

Comparison variable	Educational level	Average score	STDEV	F-test value	Significant level	Interpretation
Structural issues	Lower than secondary	3.35	0.73	0.37	0.691	-
	Secondary	3.34	0.78			
	University and higher	3.48	0.76			
Behavioral issues	Lower than secondary	3.03^B	1.01	4.26	0.015*	University and higher versus lower than secondary
	Secondary	3.15^{AB}	1.01			
	University and higher	3.69^A	1.16			
Technical issues	Lower than Secondary	3.60^B	0.80	3.56	0.030*	University and higher versus lower than secondary
	Secondary	3.68^{AB}	0.84			
	University and higher	4.08^A	0.95			
Date marketing general problems	Lower than Secondary	3.33^B	0.67	3.94	0.021*	University and higher versus lower than secondary
	Secondary	3.39^{AB}	0.70			
	University and higher	3.75^A	0.71			
Recommendations for improved date marketing efficiency	Lower than Secondary	3.59^B	0.72	4.25	0.015*	University and higher versus lower than secondary
	Secondary	3.73^{AB}	0.70			
	University and higher	4.02^A	0.59			

*Significant at 0.05 significance level   .
*Means followed with same letter do not differ significantly using Duncan's test.
Source: Study Sample, 2012

correlation analysis were utilized to investigate the significance of these variables 'effect on the following:

Types of Marketing Channels

Table9 shows a significant difference in date marketing problems, mainly through tech
nical issues between different marketing channels. In addition, recommended solution scoring varies significantly among various types of date dealers. In general, all marketing channels have higher scores for problems and suggested solutions compared to wholesalers.

Geographic Factor

Table 10 shows insignificant regional differences in satisfaction level with the performance quality of marketing services in the market, the problems of market structure, and the recommended solutions. However, significant differences do exist in the technical problems in marketing; technical problems in date marketing are significantly higher in Al-Hasa and Al-Qassim than they are in Riyadh.

Educational Level of Dealer

The date dealer's educational level causes significant variation in the marketing problems, especially in behavioral marketing and technical marketing problems. A higher education correlates with higher scores for such problems than a lower educational level. Higher educated dealers also give higher significant scoring for recommended solutions than less educated dealers (see Table 11).

The other explanatory variables, such as the extent of the

dealer concentration in date merchandising and the length of experience in date merchandising, showed insignificant variation using one-way analysis of variance.

CONCLUSIONS

Saudi Arabian date marketing system encounter many issues that hinder their efficiency and lower the competitiveness of Saudi dates at the global markets. The prevailing problems and obstacles were categorized into three major dimensions: technical, structural, and behavioral. The impact of technical problems and obstacles were found to be the highest compared to the other dimensions, with an average of 3.67 on a Likert scale, followed by structural problems with an average of 3.36 and behavioral problems with an average of 3.12.

The impact of applying the recommended solutions to overcome or minimize the impacts of the Saudi Arabian date marketing issues were found to be high and to enhance the date marketing efficiency by about 51%. They ranged between a minimum of 3.35 for encouraging competition and a maximum of 3.98 for providing affordable cold storages, with an average of 3.68. Factor analysis suggested categorizing the set of recommended date marketing problem solutions into four factors in order to improve date marketing efficiency. These four factors explain 71.02% of the variations of all of the solution items recommended.

This study also found that factors such as marketing channel types, study regions, and date marketing dealer educational level significantly impact the prevailing issues of Saudi Arabian date marketing systems, as well as the fact that these significant impacts varied among types of marketing channels, regions, and date dealer educational level.

In conclusion, the more quickly the prevailing problems of the Saudi Arabian date marketing system are addressed, the more likely the marketing efficiency and Saudi date global market competitiveness will improve.

ACKNOWLEDGMENT

The study team is very grateful to King Abdulaziz City for Science and Technology (www.kacst.edu.sa) of Saudi Arabia for its financial support and on-going encouragement. This work was funded through the Strategic Technology Program – the National Plan for Science, Technology, and Innovation (http://maarifah.kacst.edu.sa) grant # 600-32.

REFERENCES

Al-Abdulkader AM, Al-Kahtani SH, Ismail SM (2015). Marketing of dates in Saudi Arabia: issues and solution. King Abdulaziz City for Science and Technology. Saudi Arabia.

Al-Abdulkader, Ahmed M, Al-Kahtani, Safar H, Ismaiel, Sobhy M, Elhendi, Ahmed M,Saad, Ali I; Al-Amari, Yousf A, and Al-Dakhil Abdullah I. (2016) enhancing marketing efficiency of the Saudi Dates at the national and international markets. International Journal of Economics and Finance, 853-70.

Al-Kahtani, S. and El Feel, M (2006) Marketing costs for some vegetables and fruit crops in Saudi Arabia. Alexandria for the Exchange of Scientific, University of Alexandria, 27.131-148. Al-Kahtani, S, Ismail S, Aleid S, Bakri H. (2011) the prospects and possibilities of e-trade of the Saudi dates and its economic role of supporting agricultural sector. Date Palm Research Center, King Faisal University. Saudi Arabia.

Al-Kahtani Safar, AlQunabet M, Ismaiel SH,Hebaisha (2007) Agricultural marketing in the Kingdom of Saudi Arabia: existing situation, problems, and solution. Projects King Abdulaziz City for Science and Technology. Saudi Arabia.

Emana B, Atarı-Seta V, Dinssa F, Ayana A, Balemi I, and Milkessa T (2015) Characterization and assessment of vegetable production and marketing systems in the humid tropics of Ethiopia. Quarterly Journal of International Agriculture, 54 163-187.

Han Chun-mei Obstacles to Development of Marketing Channels of Agricultural Products in China and Countermeasures. Asian Agricultural Research. 3 (7): July 2011.

Hassan K, Raha S (2013). Improving the marketing system performance for fruits and vegetables in Bangladesh. Technical Report. Department of Horticulture. Bangladesh Agricultural University. 274.

Ismaiel S, Aldwais AA, Elaiwy M (2009) the use of cost function to estimate production efficiency measures for Sukkari dates in Qassim Region. The Bulletin of the Society of Saudi Agricultural Sciences, 8 28-39.

Ismaiel S, Alshehry A (2015) factors affecting the power of marketing problems of date in Alkharj Region, Saudi Arabia. The Bulletin of Saudi Agric. Science Association 2 (14):

Kriesberg M (1970). The Marketing Challenge Distributing Increased Production in Developing Nations. United States Department of Agriculture. Economic Research Service. Foreign Agricultural Economic Report (FAER).

Rust RT, Ambler T, Carpenter GS, Kumar V, Srivastava

RK (2004) Measuring marketing productivity: current knowledge and future directions. Journal of Marketing, 68 76-89.

Sheth JN, Sisodia RS, Sharma A (2002) Marketing productivity: issues and analysis. Journal of Business Research, 5(5): 349-362.

Voboril Dennis Indonesia as a Growth market: Challenges and Opportunities. United States Department of Agricultural. Agricultural Outlook Forum. 2013.

3

Dairy business value chain analysis in Lamjung district of Nepal

Thaneshwar Bhandari

Assistant Professor of Agricultural Economics, Department of Agricultural Economics, Tribhuvan University- Institute of Agriculture and Animal Science, Lamjung, Nepal.
Email: agecon.iaas.2069@gmail.com,

Among agricultural sub-sectors adopted, dairy business was one the most profitable agri-businesses in Nepal but past studies to support this statement was lacking in Lamjung district Nepal. The survey was carried out from April 2013 to January 2014 with the aim of analysing chain functions, capabilities of and support level of operational service providers, value addition, and market analysis of milk business actors. The study collected primary information from 97 respondents by using focus group discussion, key informant survey, observation and SWOT analysis. Results after using descriptive tools identified six chain functions. Estimated 33660 farming households milked 15272 tons raw milk annually but marketed only 13 percent milk and milk products in the 23 peri-urban local market-outlets through two routes: 784 tons milk fed through 7 small-scale chilling centres particularly cold chain process and 1201 tons through hot-milk base processing. Unmet 297 tons (32%) dairy products, all in processed form, were supplied from adjoining districts. The value addition analysis of cow milk showed that not only producers and processor added the largest cost share but also received the highest profit share among the succeeding agents. However, pricing and payments of dairy product were buyer-driven without making contract and no system of market sharing among the micro-actors.

Key words: Dairy business, value chains, market outlet, margins, profitable, self-life, SWOT

INTRODUCTION

According to Community Livestock Development Project (CLDP) prediction for Nepal, an annual growth rate of milk products is 10 percent against past assumption of 11 percent growth in demand. Basis of analysis is growth trend of processed fluid milk demand from formal sector, population growth and income elasticity (FAO, 2010). Same report highlighted broader corners of milk production, collection, processing, marketing and distribution functions and their stakeholders but not provided location specific interventions implemented.

One of the highly promising mill-hill districts for dairy business, by redirecting money from cities to rural areas, is Lamjung District which has significant share in national milk production and consumption because estimated 70% households of the total 42079 households keep 2-3 livestock units as integral part of the farming system in coordination of the crop and forestry sector (CBS, 2014; DLSO, 2013; CECI, 2011, and DDC, 2013). Statistics shows that buffalo population is 32% higher over cow number of 7991 in fiscal year 2011/12. About 19% cows and 42% buffaloes are being milked and their collective milk production is 11631 tons. Growth rate of improve breeds over local breeds is 113% and raising local breeds seem reducing trend by 13% in the recent year (DoLS, 2013 and DLSO,2013). Nepalese Agriculture Perspective Plan (1995-2015) has prioritised "milk production" sub-sector as one of the best suitable outcome-base commodities for mid-hill districts in Nepal (MoAD, 2009). What are the types of value chain functions and value addition activities were there by the upstream and downstream chain actors, or any policy defect on sizable milk volume production and marketing? What level of supports are gaining from public/ civil society agencies to endorse backward and forward networks? How does private sector leading the milk business? Why does a few dairies only collecting very less milk from the inhabitants irrespective of large supply from adjoining districts? What types of milk products would be profitable in such a specified market situation? No past literatures have answered these issues.

Therefore, researcher felt crucial to use value chain approach (VCA) as the best solution of sustainable milk business because VCA does not limit policy use at local level, district, country as well as links in the global level (Humphrey & Schmitz, 2001 and Humphrey,2006). Because VCA is a newly practicing market-base tool determining how business receives raw materials as input, add value to the raw materials through various processes, and sell finished products to the intended customers (GTZ, 2007; Poudel, 2008 and Keane, 2008). Since 2000s, value chain analysis has gained considerable popularity in Nepal (Banjara, 2007 and Bhandari, 2014). Value chain essentially represents enterprises where many producers and marketing companies work within their respective businesses to pursue one or more end markets (Porter, 1985 and Carmen & Demenus, 2009). Value chain participants cooperate to improve overall competitiveness of the final product, but may also be completely unaware of the linkages between their operation and other upstream or downstream participants (Kaplinksky and Morris, 2002, and Hoermann et al., 2010). Therefore, VCA encompass all of the factors of production including land, labor, capital, technology, and economic activities including input supply, production, transformation, transport, handling, marketing, and distribution necessary to create, sell, and deliver a product to an intended destination (Tchale and Keyser, 2010). The commodity chains are border-crossing value adding networks of labour and production processes whose end results are finished commodity at use (World Bank, 2012). Thus, specific objectives of the study were : i) identify dairy chain functions and capabilities of micro-actors; ii) to map micro actors and analyze opportunities and constraints under dairy sub-sector; iii) quantify economics behind value addition in actors' operations and market analysis of the milk products.

MATERIALS AND METHODS

Data source and collection methods

Rapid market appraisal (RMA) tool supported collecting in-depth facts and first hand information from the primary informants. Key methods of RMA were: focus group discussion (FGD), key informant survey (KIS), observation, case study collection and validation workshop (SNV, 2010). Checklist preparation undertook for FGD and KIS as pre-field works as per relevancy of the organization and types of service they were rendering. Then, all checklists pre-tested with the non-group members of the Kunchha village and errors were corrected in it. Potential stakeholders to be visited and its itinerary prepared with the help of staffs at District Livestock Service Office (DLSO) Lamjung Nepal. Then, all particularly dairy entrepreneurs, farmer's group and cooperatives within the cow and buffalo pocket area consulted. Consulted other actors were Model dairy Yampaphant Tanahu, Safal Dairy Pokhara, District Cooperative Union Damauli and Division Cooperative Damauli which were major business suppliers of the dairy products.

Researcher collected primary data in two seasons: flush season (June to July 2013) and lean season (January 2014) by using method of Upadhyay, Singh and Koirala (2000). In a latter case, same respondents consulted to confirm seasonal change, selling variability and quantity change. The study conducted ten FGDs: four with the farmers' group and six with primary milk producing cum selling cooperatives, altogether having 44 respondents each ranging 7 to 15 members. In a same vein, 47 Key Informant Survey (KIS) conducted by consulting: proprietors (of private dairy farm, private dairies, teashop and hotel), and managers of (line-agencies and Lamjung Chamber of Commerce). By observation method, FGD and KIS responses triangulated on investment at dairy animal-shed, counting dairy animal numbers, feed and fodder management, chilling centre, milk analysis and data-keeping system, and payment system. In addition to primary information, study also collected secondary data from published reports, journals, proceedings and web browsing of the related organization. Both shadow and market price of the labours, input and tool prices, and milk products at district level were also collected.

Analytical methods

In a beginning, cost of milk production (variable and

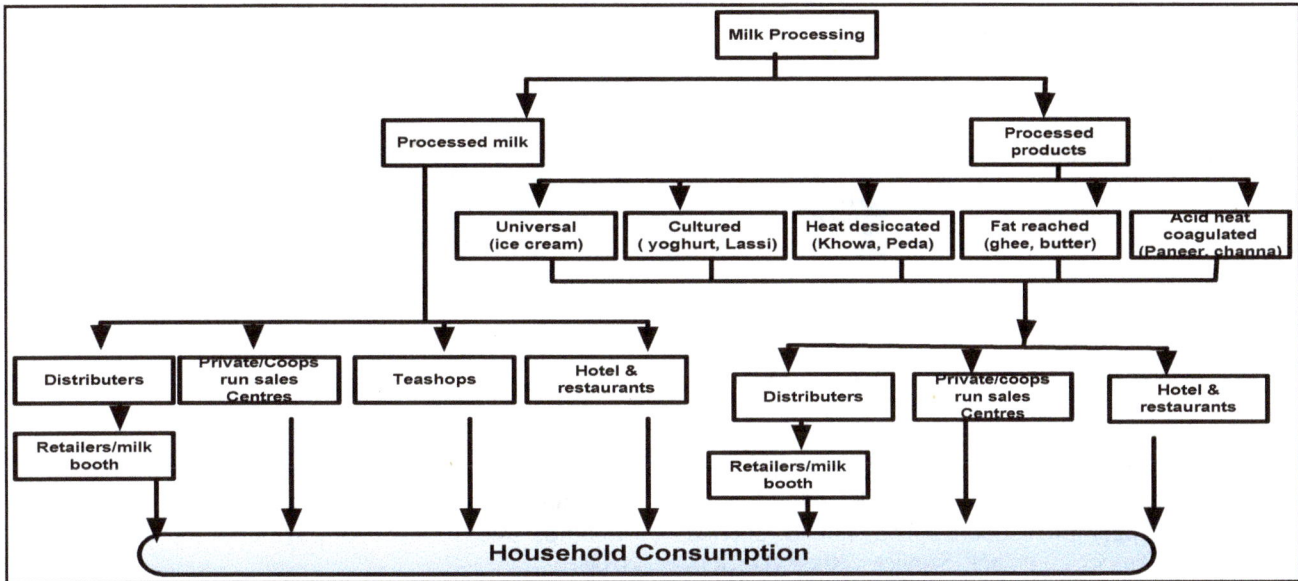

Figure 1. Types of Processed Milk and Milk Products and respective selling outlets
Source: Own drawing from field survey (2014)

depreciated fixed assets, cost of processing, unit margin, gross returns, and value addition) in various steps were analysed and averaged for micro chain actors. The basic equations of the cost function were:

$$C = f(Y),\text{-------------------------------------}(1)$$

$$C_i = f(W_{it}, T, P_f, R_{cm},)\text{----------------------}(2)$$ under long-run case (Koutsoyiannis, 1991).

where, C_i: total cost incurred by a ith household; Y = milk production as output; W_{it}: output (quantity of milk per lactation of animal), T= cost of technological option, P_f = price of factors used, R_{cm} = repair and maintenance. The fixed cost items were shed preparation, water pipes and tanks, manger, or motors, and their respective prices. Study calculated average cost of raw milk production after consulting individual farmers of different village (near to town, village and peri-urban areas), types of dairy animal (cow, buffalo or both) and number (small to medium farm raising). The study collected case studies of cost-benefit analysis of varying three entrepreneurs' farm raising buffalo alone, cow alone, and mixed of both cows and buffaloes in order to compute fixed costs, operational costs, net-income before tax, labor cost and per litre cost of production. Tchale & Keyser, 2010). Using Zimmerer and Scarborough (2008), break-even analyses of these farms were calculated in a unit basis by using following formula:

Break-even volume:

$$= \frac{Total\ fixed\ costs}{Sales\ price\ per\ unit - Variable\ cost\ per\ unit}\text{---}(3)$$

Study calculated comparative calculations related to value addition while preparing cold chain-base or hot milk-base diversified products (see figure 1). Further, marketing analysis tools used to determine dairy inputs, and market schedule of milk products in flush season (15th July to 15th December) and lean season (16th December to 14th July). Types of milk suppliers, their seasonal capacity and accordingly types of consumption outlets as well as import and export volume as per market outlet were analysed by considering demand and supply situation. With incorporating linkage, prices, dairy products, and actors, a complete chain map portrayed into the value chain map (see figure 3).

All collected information were presented to the participants from herd raisers, service providers and traders during the validation workshop. In panel discussion through SWOT analysis method, these participants outlined district level market chains, and backward and forward linkages, which were basis of drawing value chain map.

RESULTS

Value chain functions of the micro actors

In order to reflect objective one of the study, six value chain functions of the actors identified that were input supplier, raw milk producer, collector, processor, trader and consumer (see figure 3).

Input supplier: The horizontal level of input suppliers were dairy resource centres and cattle and buffalo

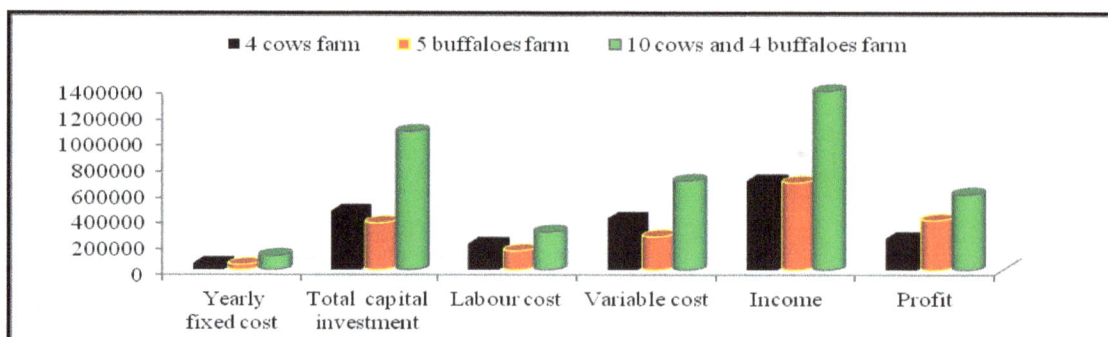

Figure 2: Cost and Returns of Viable Cow and Buffalo Farms
Source: Own calculation from field survey (2014)

pockets, feed suppliers, pet-vet/agro-vet shopkeepers, artificial inseminator (AI), Village Animal Health Workers (VAHWs), dairy animal keepers. The public service providers were District Livestock Service Office (DLSO), its Service and Sub Service Centres, and a few NGOs. The DLSO identified and intensified support to nine cow raising pockets and six buffalo raising pockets. The fodder and forage resource centres were Chiti, Khudi, Bagre, Dhamilikuwa, Chakratirtha and Archalbot VDC. The study found 36 pet-vet shops retailing medicines, minerals, vitamins, minor tools, equipments and forage seeds equals to 5 million turn-over per year. Including officials at DLSO, the trained 18 inseminators and 145 VAHWs provided AI services to cow and buffalo, castration and basic treatment services at village level. There were four feed suppliers that transacted annually 20 tons feeds brought from Rupandehi and Chitwan. Within resource centres, there were three local nursery owners selling nutritious fodder seedlings and grass sets.

Raw milk producer (RMP): Dairy animal raisers, around 33600 households, were producing raw milk equals to 3857 tons and 10927 tons milk from 21% cows and it's just doubled she-buffaloes they milked, respectively. Share of improved breeds, especially 5337 buffalo and 816 cattle on milk production, was 65% on total buffalo milk and 80 % on cow milk, respectively.

Milk collector: The milk producing cooperatives (MPC), agricultural cooperatives and private dairies, a few in numbers, were collecting milk from RMPs through the establishment of systematic collection centres (CCs) or temporary arrangements. These institutions provided "Pass Book" for all RMP in order to record sell amount, SNF, fat value and temperature of collected milk. DLSO, PACT, and World Vision were partially supporting in the establishment of 12 CCs in the district. As soon as RMP milked their animals, they usually collected raw milk at the adjoining CCs; each collected 50-270 litre milk daily in the morning time. Recruited from milk buyer's side, the

staff of the CC undertook fat and SNF content sample test and kept records of incoming and outgoing milk volume, and respective payments of the milk sellers (RMP).

Milk chilling was not counted as a specific ladder in the study area but it was part of storage longing process of 1350 tons milk carrying into practice of cold chain by six small-scale chilling vats operators of MPCs (Dudheshower, Janta, Janaunnati, Himal, Annapurna, and Agri-business Promotion), one group managed (Manakamana farmers' group) and 2 private firm-led (Kusal and Annapurna) dairies. Collected milk after milk testing went immediately into chilling process managed by the MPCs or private dairies (Annapurna, Marshyandi and Kusal). Downing temperature to 4-5°celcius prolongs keeping quality of milk upto 72 hours so that chilled milk could use safely in a long distance travelling.

Milk processers: Organised sector handled 729 tons milk in the form of cold chain processing. As shown in figure 1, study identified six-type of cold to hot-milk based processed products: cultured products (curd or yogurt, Lassi), heat desiccated products (Khoa, Rabri), acid heat coagulated products (Chhanna, paneer), fat rich products (cream, butter, ghee) milk based pudding (Khir, Haluwa) and universal products (ice-cream, cheese). The whole milk was transformed to standard milk by maintaining 3% fat. The unorganised sector involved dominantly at hot milk-base processing where 726 tons milk (by the teashops) and individual farmers and 927 tons by hotels/restaurants consumed from estimated milk 1653 tons per annum. Further discussions showed that 30% milk products (out of 726 tons) were directly sold in the form of local processing. Simple processed 12245 tons milk was consumed at home. Study also estimated nearly 2487 tons milk for hot-chain milk products, 9956 tons of milk semi-fermented for Dhahi (of volume 8960 tons).

Traders: The milk traders performed formal wholesaling

Table 1. Value addition per litre of cow milk

Value addition indicators	Producer	Collector	Processor	Distributor	Total
Production cost/buying price	23.0	32.8	40.3	50.0	
Transport cost	0.0	1.2	2.5	0.0	3.7
Testing/ administrative cost	0.0	2.5	0.5	0.0	3.0
Processing cum chilling cost	0.0	1.0	2.2	0.3	3.5
Rent/ labour cost	3.0	0.3	1.3	0.5	5.0
Packaging and labeling	0.0	0.0	1.0	0.0	0.0
Total cost	26.0	37.8	47.9	50.8	
Added cost (in NRs and %)	26.0 (66.2)	5.0 (12.6)	7.5 (19.1)	0.8 (2.0)	39.3 (100)
Sale price (NRs)	33.0	40.3	50.0	52.0	
Sale byproduct (cream 1%)	0.0	0.0	4.0	0.0	4.0
Profit in NRs (sales price -cost) and	7.0 (21.3)	2.6 (6.3)	6.1 (12.3)	1.2 (2.3)	16.9
Added profit (%)	41.6	15.1	36.3	7.1	100

Note: Figures in parenthesis shows percentage
Unit:- Nepalese Rupee (NRs): 1 US$ = 95 NRs
Source: Own estimates from field survey (2014)

and retailing as well as informal trading. Mostly cooperatives and few suppliers (Joshi and Hulaki) practiced wholesaling by selling dairy products relatively 6 to 12% cheaper than the retailers. Unlike it, study found majority of the milk retailing actors had no forward contract or cordial business relationship.

Consumers: Study recognized 23 peri-urban areas (such as Besishahar, Bhoteodar, Sundarbazaar, Bhorletar, Udipur, and so on) as dairy product consumption outlets. General estimation of District Hotel Association and Chamber of Commerce and Industries (CCI) revealed that hotels, fast-food preparing restaurant, tea shops cum sweet-shops were major consumption outlets of processed products (Table 3 and figure 3).

Chain supporters in the dairy-sectors

Of the identified micro-actors getting services from the market chain supporters (meso-level and a few macro-level) are shown at right side of figure 3. The meso-level chain supporters were: Livestock Market Management Directorate (LMMD), Regional Livestock Directorate (RLD), Regional Livestock Service Training Centre (RLSTC), Regional Veterinary Lab, Division Cooperative Office, District Agriculture Coordination Committee, Commercial Banks, District Livestock Service Office (DLSO), District Development Committee (DDC), Cottage and Small Industries Committee, Institute of Agriculture and Animal Science (IAAS) Lamjung, Chamber of Commerce and Industries (CCI), Village Development Committee (VDC), Small Farmer Development Cooperatives Ltd, 41 committees of cow/buffaloes

pocket, 19 Dairy Farmers Coordination Committees, District Cooperative Union (DCU), 49 VAHWs, 9 AI trainers, Chess Nepal, World Vision-Area Development Program, and Private Dairy Association Tahanu, Chitwan and Kaski. Likewise, major macro-level chain supporters especially supporting in policy formulation and budgetary provision were: Ministry of Agricultural Development (MoAD), Ministry of Finance (MoF), Ministry of Supply and Commerce (MoSC), Department of Livestock Service (DoLS), Dairy Development Board (DDB), Nepal Agricultural Research Council (NARC), Project of Agricultural Commercialization and Trade, National Milk Development Board (NMDP), National Dairy Cooperative Union, National Animal Breeding Centre Pokhara, Federation of Chamber of Commerce and Industry (FNCCI), IAAS Rampur, and Central Banks.

Economic analysis of dairy business

Study used Equation (2) for computing cost of milk production, and returns with value addition generated by the micro actors. The further computation was income distribution among the value chain actors in terms of value share, margin, and cost of production of dairy products.

Cost of production and returns: Figure 2 depicts average fixed cost, recurrent cost, net-income before tax, labor cost, and per liter cost of production of three farm sizes.

Unit: Nepalese Rupee (NRs): 1 US$ = 95 NRs

Table 2. Value addition on one litre milk for different products
Unit: - Nepalese Rupee (NRs): 1 US$ = 95 NRs

Cost categories	Milk	Hot milk	Tea	Ice-cream	*Khoa*	Paneer	Curd or yoghurt
Market price of unit	50.0	20.0	10.0	550.0	450.0	475.0	70.0
One litre equals to	1	4	10	0.7	0.2	0.2	0.9
Buying price	40.0	80.0	60.0	366.7	75.0	73.1	63.0
Firewood for boiling	-	3.5	3.5	7.0	3.5	3.0	1.0
Labour	-	0.5	0.5	15.0	7.0	5.0	0.5
Citric acid	-			0.0		2.0	
house rent	0.3	0.2	0.0	1.0	1.0	1.0	0.5
Other materials	-		10.0	58.0	0.5	2.0	0.3
Utensil and preservatives	1.0	2.0	1.0	20.0	2.5	2.0	1.2
Total value addition	1.3	6.0	15.0	100.0	13.5	14.0	3.0
Total cost	41.3	46.0	55.0	140.0	53.5	54.0	43.0
Profit	8.8	34.0	45.0	226.7	21.5	19.1	20.0
Compare profit		288.6	414.2	2490.5	145.7	118.0	128.6

Source: Own estimates from field survey (2014)

Yearly deprecation of a buffalo farm was lower than other firms. Per animal fixed cost and recurrent cost was 99000 and Rs 71000, respectively. Comparing labour cost of the three farms, it showed 50%, 56% and 42%, respectively for the three farms. Other efficiency indicator was cost of production of one litre milk which was NRs 22.31 for cow only farm, followed to mixed firm by NRs 25.46 and NRs 27.22 for buffalo only farm. Discussion revealed that 8-month and 10-month long was average lactation period of buffalo and cow, respectively. The period to come into break-even point of these three farms were 1.65 year, 1.12 years and 2.12 year, respectively.

Value addition analysis of cow milk

Only researcher took cow milk into value addition analysis is shown in table 1.

Producer's cost of production was NRs 23.0 and additional cost of labor to bring into collection centre was NRs 3.0. Thus, net income of selling litre milk to nearest collection centre (or market outlet) was NRs 7 after selling it by NRs 33. Subsequently collector added value of NRs 5.0 on top of NRs 33.0 and sold it to processor by receiving NRs 40.3. The processor used to NRs 7.5 additional cost and sold it at NRs 50.0 to distributers by making profit of NRs 6.1. The 1% cream as byproduct was taken into calculation because almost all processors churned additional fat of the whole milk. Last selling points of the distributer were hotels, tea shops, individual households etc by receiving NRs 52 for standard milk and NRs 58 for processed whole milk.

Cost and benefit of product diversification

Table 2 depicts types of milk product, cost addition while preparing diversified products, retail market price at local level and net profit while converting one litre raw milk. Culturally and traditionally popular Khoa (milk cheese), curd (or yoghurt) and paneer were manufactured by almost all processors by adding value. *Khoa* was prepared from whole milk boiling in an open skillet for long hours in the ratio of five-litre milk equals 1 kilogram *Khoa*. Likewise, paneer was manufactured by adding fixed proportion of citric acid (or lemon juice) into hot milk, separated solid whey from greenery water by regular stirring it, prepared a different size paneer after using cooled whey through using cheesecloth in a strainer, rinsing the curds with the fresh water, and squeezing out moisture from the curds. Another product curd or yoghurt was processed from boiled milk, transferred the milk to the container in order to set the curds in, and added pre-cultured curd as a starter. A bacteria called *lactobacilus* converts lactose into lactic acid thus making it sour and turns it all to curd.

Market analysis

Input marketing: FGD on backward linkage revealed that buying production tools, live animal, dairy equipment, feed, forage and fodder seeds were dependent to external market. Importantly, there was no outlet for buying and selling dairy animals. Per buffalo price was NRs 45000- 65,000 and improve cow could buy by NRs 55000-70000. Strangely, expensiveness was also

Table 3. Supply and demand situation of dairy products
Unit: - Metric tons

Supply schedule				Demand schedule					
Means of Supply	Flush	Lean	Total	Means of consumption	Estimated consumption		Total	% of use	
					Flush	Lean			
Private dairies (Lamjung base)	74	88	162						
Model dairy Tanahu	60	74	134	Ice-cream industry	4	4	8	0.05	
Safal dairy	150	147	297	Tea & sweet shops	323	240	563	3.71	
MPC	287	243	529	Restaurant & hotel	392	493	885	5.84	
Farmer direct sell	335	392	727	Export to Damauli	23	17	40	0.26	
Home consume	6551	6771	13322	Home consume	6551	6771	13322	87.82	
Total supply	7456	7714	15170	Processing damage	40.09	32.97	73.06	0.48	
Market sell	905	943	1848	Demand in towns	123.36	156	279.36	1.84	
Local sale (%)	12	12	24	Total	7456	7714	15170	100	
Import (%)	17	16	32						

Source: Own estimates from field survey (2014)

observed local buffaloes too whose live price was at least NRs 35000. Even older buffalo could sell at NRs 17000-20000 for meat purpose. The study found high demand of dairy equipments, lab materials and chemicals.

Output marketing: Study calculated supply and demand schedule of output marketing through various market outlets of diverse dairy products during flush and lean season are presented in table 3. Annual supply of the dairy products was 13322 mt but sold amount was 24% (1848 mt). The unmet 32% quantity, brought from Kaski and Tanahu District, was especially sold to ice-cream industry, town dwellers, and restaurants and hotels.

Pricing and payment systems: Pricing of dairy animal was oxen system in the floor between the buyer and seller. However, pricing of dairy products was relatively buyer-dependent. Government was involved in fixing TS base pricing where fat and SNF value were priced NRs 3.4- 4.0 per gram and NRs 2.4- 2.6 per gram, respectively. Maximum rate was officered during lean season while minimum rate for flush season. By computation, one litre cow milk having fat 4 % and SNF 8.1% would price 3.4* 4+ 2.4* 8.1= NRs 33.04 while a litre of buffalo milk having fat 6% and SNF 8.5% had price 3.4* 6+2.4*8.5= NRs 40.8. Milk testing staffs reported changing milk quality each day, could change also value of milk at buyer's gate. The top of commission for forward buyers was 26 % of the total solid collected i.e. 4+8.1=12.1*26% = NRs 3.15 per litre milk sale. Some experienced cooperatives did price negotiations prior to milk collection. However, pricing of other dairy products

was market based. Study found every fortnight was due duration of payments of the milk buyers.

Drawing value chain map

Using Meger-Stamer and Waltring (2007) idea, a schematic map has been shown in a linear sketch in figure 3 beginning from raw material used in production, production to transformation stages into finally at end market outlets (consumers). Same figure also shows data and information on operational service providers, product and price flow, volume, as well as kinds of service provided by the service enablers.

DISCUSSIONS

Discussions on chain functions revealed that most of the agents involved in multiple functions: production, collection, processing and trading. For example some processors were also wholesaling cum retailing their dairy products through using various market outlets. Majority of the growers fed cereal by-products rich feeds, forage and ration to their dairy animals without caring animal size, productivity and nutritional requirements. Almost all respondents neither built technically suitable shed nor arranged basic requirements of animal keeping and feeding management irrespective of estimation of 33600 farm households investing around 8 billions in dairy business at least NRs 2,50,000 each and employing a job for a efficient person-day. Might be the cause of these limitations, the economic life of the dairy animal

Figure 3. Milk Sub-sector Value Chain Map in Lamjung District

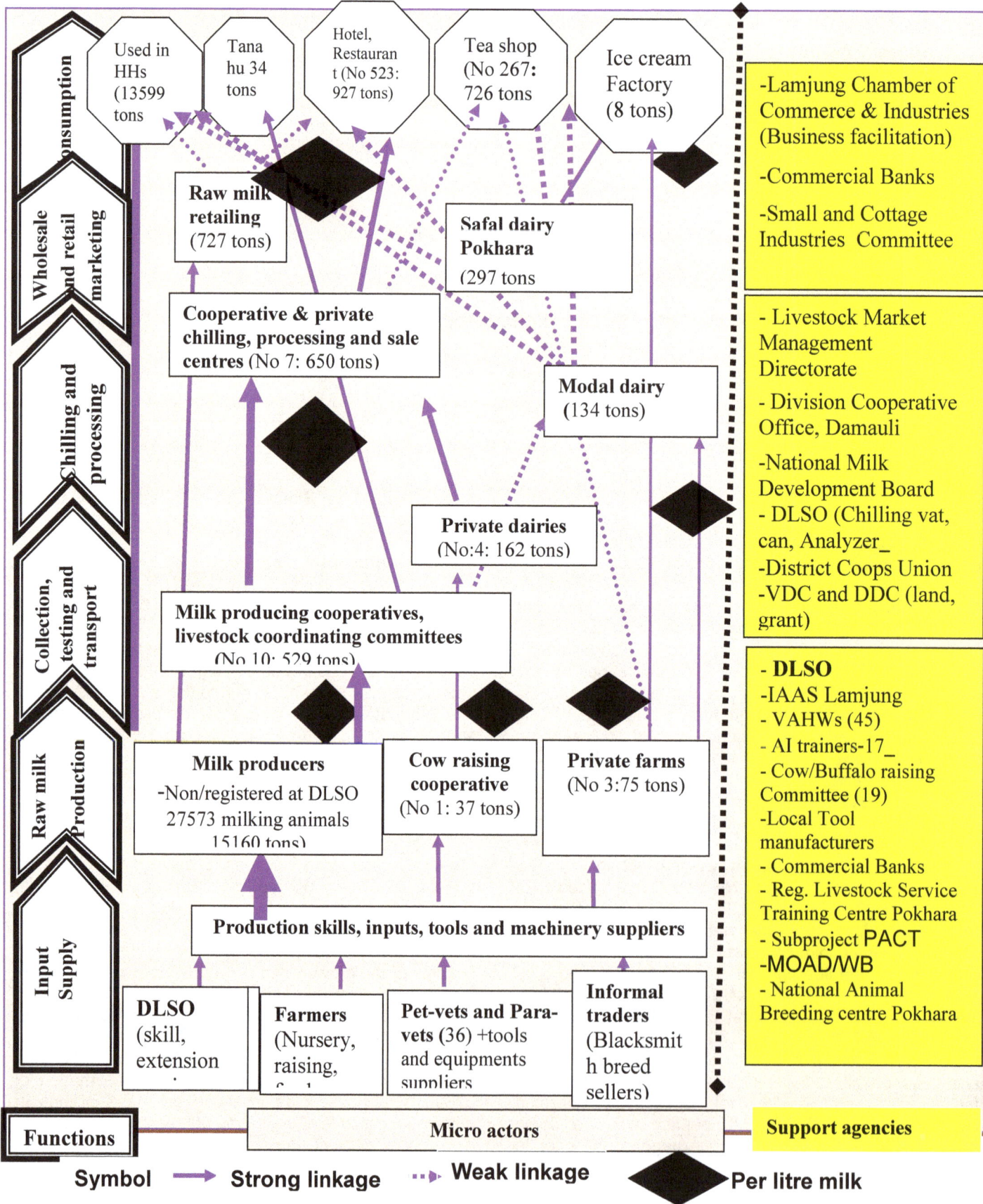

Figure 3. Milk Sub-sector Value Chain Map in Lamjung District

and milk quantity were substantially reducing over the subsequent lactations. Estimate showed local cow and buffalo had one-seventh and one-half level of milking capacity in comparison to their respective improved breeds, respectively during their lactation. Except a few local origins, most of the dairy tools cum equipments were importables of India and China which were highly expensive and technical skill-base. Also, there was no repairing person for dairy equipments. Majority of the pet-vet keepers and farmers mentioned food and mouth disease, mastitis, liver fluke, indigestion as economically important diseases. Study identified very less productivity of local breeds than that these had potential yield. It meant local breeds had nominal share on overall milk production. Unlike FAO study, far better position was reported in case average milk productivity of cow and buffalo because milk production per cow in the Hills was 435 litres and it was 812 litres for buffalo (FAO, 2010). Average quantity of milk of cow and buffalo per lactation was 534 litres and 537 litres, respectively, which was far behind the theoretical performance of the improved dairy animal (Broadway and Broadway, 2008). Unlike it, per day productivity of the improved cow was 10-12 litre milk upto 10 months while local breeds had 1.2 litres per day in a nine month's lactation. Similarly, it was 5-6 litres for pure Murrah buffalo and 3-4 litre for cross Murrah in their seven-month long lactation. Farmers reported longer lactation period for local buffaloes in case of unsuccessful conception. The discussion prevailed that milch or pregnant breed were highly expensive and no readily availability.

Whenever the raw milk brought into collection process, four issues repeatedly came into deterioration of whole milk. Some collection centres locating two to more hours' travel distance faced milk coagulation or souring because of lacking chilling facility especially collected it from newly established MPCs. Some MPCs/CCs were collecting very minimum milk quantity than that adjusting operating cost of chilling vat. Second issue reported by MPCs was malpractice of the sellers by selling: stale milk, immature (colostrums) or milk of animal treatment period. Third issue was frequent power cut-off in the chilling place that was disfunctioning some chilling vats. Some milk chilling operators faced electricity problem in lieu of delaying in transform two phase to three phased system. Using diesel plus repair and maintenance cost of chilling vat had been increased per litre chilling cost upto 11-15%. Fourth issue was not seriously following code of practice of routine sanitation of milk plant, utensils, and milking animal. From individual farmers to majority of the CC used dirty aluminum cans, jars or plastic gallon while bringing or milking, keeping milk, curds because of not having practice or awareness of using chilled Lorries in transportation.

Milk chilling was considered major function without forming a specific ladder in the study area but only 9 % raw milk of the total sale came into cold chain through six small scale chilling vats operated under MPCs and 2 private dairies. Other low cost option was using deep freeze or simple refrigerator. The processing capacity of organised sector (Such as private dairies and dairy cooperatives) was growing year after year and rate was higher for private dairies than MPCs. However, both actors underutilised (65-70% only) chilling and processing capacity. It was also revealed that the some actors were using relatively high-tech chilled to frozen products having long self-life (like ice cream, sweets and cheese). However, the size of the universal products market was disproportionately small in comparison to overall milk product consumed in the local market. Annually 11 tons fat-rich paneer was consumed inside the district with the daily supply made from local manufacturers. The cream demand of two ice-cream factories hardly fulfilled from internal production so was importing from other district. Unorganised sector was manufacturing and retailing temperature ambient low shelf-life products through using simple processing. For example, raw milk producers directly retailed dairy products to hotels, teashops or sweetshops without assuring quality.

Yearly fluctuations in milk demand and supply during lean to flush season had adjustment problem of the processors in lieu of absence of bulk processing plant, no batch pasteurization practice, monopolistic competition in getting milk, and poor packaging and labelling practices of manufactured products, which ultimately increased distribution costs (see table 1). Study reported increased consumption of income elastic milk products in town dwellers. Drinking boiled milk in a morning or with lunch as well as dinner, using milk tea/coffee was traditional way of milk consumption in the study area. Almost 82% produced milk was consumed at home by using simple processing i.e just boiling and consuming or churning 2-3 days stale curd into buttermilk, butter and ghee under hot milk-base product making. This study estimated roughly 80 litres equivalent per capita milk consumption, which was relatively higher than national average of 57 litres (DLSO, 2013). It meant though own production or importing from other districts, per capita consumption was relatively higher than the national average (DLSO, 2013). The milk suppliers did not address quality concern of the consumer in the past but the Department of Food and Quality Control (DFTQ) sanctioned one milk company as defaulters in the presence of coliform in dairy products of last years' supply. Other points of distraction of keeping livestock was gradually increased over the years because of no younger's interest on farming irrespective of rapidly growing demand of dairy products in the peri-urban, urban and home-stay (because of yearly approximately 30 thousand tourist arrived in a district), and increasing labour wage over the year.

As a major input provider of the public sector office, the DLSO was also supporting fodder seeds, grass-sets, tools, and veterinary services in a subsidized rate under extension support programme. DLSO had 18 high to junior level technicians for providing extension services through district office and 11 local service centres by annually allocating 30% budget at dairy sector of total annual programme. Considering high price of milch breed cow, DLSO Lamjung had nominal price support to yearly 40 improved cattle. Other public service providers and non-government organization (NGO) had joint programme of providing very nominal production inputs. Only one bank invest in agriculture was Agricultural Development Bank Ltd. Beshisahar, which only invested two millions loans in dairy sub-sector in the study area along with nominal support of animal insurance scheme.

Figure 1 depicts three cases: raising buffalo alone in Sundarbazaar VDC-5, cow-alone farm in Bhalayakharka-2 and mixed of cows and buffaloes in Sundarbazaar-4. Discussions on average investment revealed that farmers used varying rate of current, working and long-term assets. Of the calculation of total cost investment, types of animal raising was also one of the causes of increase or decrease cost. For example dairy cattle as more efficient milk producer over milch buffalo by 30 % because of shorter calving interval and longer lactation length (Upadhya et al, 2000). Unlike the result, this study estimated 1.12 year i.e. two-lactation as pay-back period for a buffalo only farm (see figure 1) because of selling milk in higher price. Value addition of a litre cow milk analysis shown in table 2 revealed that cost price and market price spread was NRs 10 per litre at farm-get but difference was 3 times higher for buffalo milk (DLSO, 2013).

Particularly costs addition and generating profit margins in each ladder had special meaning in value chain analysis. Profit in terms of sale price was the highest for milk suppliers (21%) followed by the processors (12%) and collectors (6%). Table 2 depicts total profit generation NRs 17 by adding total cost NRs 39.3 (i.e. NRs 23.0 production and NRs 16.2 marketing cost). The largest profit share was calculated for the raw milk producer (by 42%) by investing added cost of 66 %, followed by processor. Of the total 15272 tons milk produced, only 4.5% milk was handled by the collectors (3.5% from cooperatives and 1% by private dairies). Considering that statistic, farm-get revenue of the raw milk producer was Rs 504 millions (multiplied farm get

price of Rs 33.0). The collectors could receive NRs 27.88 million incomes and NRs 1.8 million net profit, organized processor handled 783 tons milk to earn NRs 39.15 millions income and NRs 4.8 million net income. Similar margins also reported by FAO (2010). PACT (2012) also reported the highest share to the raw milk producers in far-west districts. By analogy, producers and processors ultimately benefitted from the retail price paid by the consumers. Using economics of scale, this analysis also confirmed that cost of producing and processing milk would be substantially lower in case increasing firm size. Results presented in table 3 revealed that manufacturing seasonal item (like ice-cream) was highly profitable business followed by selling milk tea and hot milk in the same vain, Khoa and paneer were mild profitable items but making sweets from Khoa was highly profitable in comparison to selling standalone. These results were also supported in Indian context that diversified milk products had much more benefit than selling fluid milk (Export Victoria, 2009).

Table 3 depicts milk supply and demand from market outlets in two seasons: flush season (July 15 to December 15) and lean season (Dec 16-Jul 14). Additional unmet dairy products demand than that marketed 1848 tons was 32 % that was imported from Tanahu and Kaski district. Major dairy products consumption outlets were: household consumption (87.82%), 267 tea cum/or sweet shops used 563 tons (3.7%), 523 numbers larger hotels, restaurants and home-stay used 885 tons (5.8%), individual households in peri-urban area used 2% milk, 2 ice-cream factories used 8 tons, 40 tons raw milk exported to Tahahu district (via Damauli Cooperatives), and 0.5% losses. Standard milk, raw milk sale and traditional subcontinent products such as Khoa, paneer, curd and ghee dominated local market sale but dairy product market was growing by 5 % annually because of rapid urbanization and increased income of those usual consumers. However, maintaining coordinated supply chain of such products was challenging job of current suppliers because of above said issues.

Price fixation of dairy animal was based on lactation age, average milked quantity, udder quality, lactation length, behaviour, body structure and colour. Generally applied practice was selling breeds farmer to farmer especially within or nearby village. The fat and SNF value, which were applied throughout the country, were major factors of fixing fluid milk price (DDC, 2013 & PACT, 2012). Suspecting the milk adulteration because of fat and SNF value, however, FAO (2010) recommended cost of milk production and price inflation Although private dairy association (PDA) was not formulating in the study district, but adjoining district Tanahu and Pokhara had dominant role of it on fixing milk product prices (DLSO, 2013). Unlike it, bargaining on milk pricing was adopted for unorganised sectors. There was no bargain in milk quality but price received of each dairy product had 20-30% higher at direct sell.

CONCLUSIONS

Input supply, milk production, collection and testing, processing, trade, and consumption were key value chain functions. Some ladders were under-specified because of multiple tasks conducted by the same agent. Mainly keeping dairy animals under traditional shed by feeding under-balanced feed were key factors of poor productivity of the breeds. Analysis confirmed that cost of producing and processing milk could lower if working under economics of scale. There were very few chain supporters substantially support value chain financing, however, 3 private dairy animal farms, 4 private dairies and 12 collection centres, 7 cooperatives and cow and farmers group at buffalo pocket centres were doing novel works on dairy business commercialization. In addition to its annual production of 15272 metric tons, around 32% milk was imported from adjoining districts to meet the growing demand of milk and milk product irrespective of outgoing 34 tons raw milk in Tanahu. Of the total milk marketed, processors handled only 729 tons milk for preparing low self- life products. The major consumption outlets of dairy products were hotels, teashops, town dwellers, ice cream factories and export Tanahu districts. Study also concluded inelastic demand and supply of local live breed but income elastic market demand of cross/improved one.

Cost addition and generating profit margins in each ladder had special meaning in value chain analysis, which concluded that value capturing chain agents were raw milk producers and processors. However, pricing and payments of milk and milk product were buyer-driven without making contract and no system of market sharing among the micro-actors. No endorsement of dairy business plan, no strengthening backward linkage support (improved breed supply, modern equipments, and AI services), not caring scientific shed and feed management at producer level, not establishing large capacity milk processing plant within the district, no skill transfer for preparing diversified universal dairy products, and including other poor performance of secondary and tertiary marketing functions were major weakness of poor performance of the dairy business. Thus, new areas of investment are suggested to concerned operational service providers to boom share of dairy-sub sector in the national economy. Increased production by improving performance of existing animal (Feed, AI, management)

could easily substitute district milk demand and also can supply excess quantity.

ACKNOWLEDGEMENTS

Research Division at Tribhuvan University was highly grateful for granting competitive faculty research fund for these nobel results. The author would like to thank for those concerned respondents and stakeholders giving valuable information. Great thanks to Research Assistants: Amrit B.K., Anuja Rijal, Bhawana Bhattrai, Kumar Shrestha, Nabin Sedain and Sita Khanal for collecting parts of primary information. Finally, the author would like special thanks to all unanimous peer reviewers and editors of this article.

REFERENCES

Banjara G (2007). Handmade paper in Nepal: Upgrading value chain approach. GTZ Nepal (2nd edn), 1-24.

Bhandari T. (2014). Analyzing dairy sub-sector value chain in Far-Western terai districts of Nepal. J. of Inst. of Agr. and Anim. Scie. Rampur Campus, Nepal.

Broadway A, Broadway A. (2002). Marketing milk and milk products. In: Textbook of agribusiness management, Kalyani Publishers New Delhi.

Carmen L, Demenus W (2009). Value chains promotion guideline. Salvadorean Private Actors GTZ I: 7-15.

Central Bureau of Statistics (CBS) (2014). Population census of Nepal, Kathmandu Nepal.

Canadian Centre for International Studies and Cooperation (CECI) (2011) Dairy cooperatives in Nepal: Big changes for small communities. Final progress report, Kathmandu Nepal.

Christopher M (2005). Logistics and supply chain management: Creating value-adding networks. Pearson Education Limited, (3rd edn), pp 12-14.

Dairy Development Corporation (DDC) (2013). Annual progress report fiscal year 2012/13, Kathmandu, Nepal.

District Livestock Service Office (2014). Annual planning and progress report 2070/2071, Lamjung Nepal.

District Livestock Service Office (2013). Annual progress report 2012/13. Lamjung Nepal.

Department of Livestock Service (2013). Annual progress report 2011/2012, Lalitpur Nepal.

Export Victoria (2009). Market opportunities in Indian dairy value chain: A research study for the Victorian Government Business Office-India.

Food and Agricultural Organization (2010). Dairy sector study in Nepal, Kathmandu.

German Technical Cooperation (GTZ) (2007). Value links manual: The methodology of value chains promotion, German Technical Cooperation, 1st Edition, Eschborn, Germany.

Hoermann B, Choudhary D, Choudhury D, Kollmair M (2010). Integrated value chain development as a tool for poverty alleviation in rural mountain areas: An analytical and strategic framework. ICIMOD, Kathmandu.

Humphrey J. (2006). Global value chains in the agri-food sector. UNIDO, Vienna working paper: pp 13-52.

Humphrey J, Schmitz DH (2001). Governance in global value chains. Institute of Development Studies. IDS Bulletin 32(3), pp.1-5.

Kaplinksky R, Morris M (2002). A handbook for value chain research. Inst. of Dev. Studies. http://rmportal.net/library/content/frame/handbook-for-value-chain-research.pdf. Access on March 07, 2010.

Keane J (2008). A new approach to global value chain analysis. Overseas Development Institute, Working Paper (293), pp 2-11.

Koutsoyiannis A (1991). A modern microeconomics. 2nd Edition, Palgrave Macmillan

Meger-Stamer J, Waltring F (2007). Linking value chain and the making markets works better for the poor concept. GTZ, Online access January 23, 2013.

Ministry of Agricultural Development (2009). Collection of agriculture related policies, acts, laws and bylaws". Kathmandu Nepal.

Project for Agricultural Commercialization and Trade (2012). Value chain status of dairy in Far-Western Development region. MoAD, Kathmandu Nepal.

Poudel KL (2008). Total value chain management. In: Agribusiness Management. Himalayan College of Agricultural Science and Technology Kathmandu, pp. 291-297.

Pourter M (1985). Competitive advantage: Creating and sustaining superior performance. Accessed October 07, 2012.

Netherlands Development Organization (2010). Cardamom value chain analysis in the eastern Nepal.SNV Lalitpur, Nepal

Tchale H, Keyser J (2010). Quantitative value chain analysis: An application to Malawi. The World Bank Africa Region Policy Research Working Paper 5242.

Upadhyay R, Singh S, Koirala G (2000). A policy review on milk holidays. Research report series (44). Winrock International Kathmandu.

World Bank (2012). Facilitating smallholders' access to modern marketing chains. In: Agricultural Innovation Systems: An Investment Sourcebook. http://www.wb.org/publications. Online access January 23, 2013, pp. 52-58.

Zimmerer WZ, Scarborugh NM (2008). Essentials of Entrepreneurship and Small Business Management. 5th Edition, Pearson Education, Inc Publishing USA.

4

Economic Analysis of Poultry Egg Marketing in Oredo and Egor Local Government Areas of Edo State, Nigeria

*Oluwatuyi Toyin Bukola[1], Eronmwon Iyore[2], Emokpae Osayi Precious[3]

[1,2,3]Department of Agricultural Economics and Extension services, Faculty of Agriculture, University of Benin, Benin City, Nigeria

The study analyzed the marketing of Poultry eggs in Oredo and Egor local government area of Edo state. It specifically described the socio-economic characteristics of the poultry egg marketers and determined the market conduct, performance of the egg market, marketing margin, profitability, effect of marketing cost on the magnitude of the marketing margin and the constraints that affected egg marketers in the study area. Structured questionnaire was randomly administered to 100 respondents in 4 purposively selected egg markets based on egg market concentration in order to gather relevant data. Data gotten were analyzed using descriptive statistics, budgetary techniques and the ordinary least square model. The result indicated that egg marketing in the area was dominated by females who were mostly married and had experience in poultry egg marketing. Results from the study also indicated a fairly good market performance as market margin per crate of egg and market efficiency were fairly good. The marketers were exploitative in setting their prices as the mark-up took a greater share of the total margin. The BCR indicated that the business was viable. The result of the regression analysis showed that marketing costs (transportation, storage, market levy and tax) greatly influenced the marketing margin realized from poultry egg marketing in the study area. Credit and Loan facilities should be made easily accessible to the marketers as they complained about the lack of accessibility to loan and credit facilities.

Keywords: Poultry Egg, Marketing Margin, Market conduct and performance, Profitability.

INTRODUCTION

Poultry products provide animal protein which is essential to mankind, their meat and eggs provide a suitable substitute to other sources of animal protein (Olukosi et al2007). Poultry products are consumed all over the world and the consumption has over the last decade increased across the globe. Their relatively low cost of production, rapid growth rate, high nutritional value and the introduction of many new processed products have been attributed to the recorded increase in consumption (Barbut, 2001). According to Oji and Chukwuma (2007), the poultry goes a long way in providing animal protein for the populace, because it yields quick returns and provides meat and eggs in a very short time. In Nigeria, egg production increased so dramatically that the west regional government started an egg marketing scheme in 1962 to mitigate a glut in the market (David West, 2008).

The egg as a major product of poultry is one of the most nutritious and complete food known to man. Being the cheapest per unit source of animal protein, eggs are more readily affordable by the populace than other sources of animal protein (Abanikannda et al, 2007). According to Esingmer (1991) and Banerjee (1992) the poultry eggs nearly approach a perfect balance of all food nutrients. The yolk and albumen contain 17.5 and 10% protein by weight, respectively. Egg marketing would be highly beneficial to the society as it would eliminate the shortage in the supply of poultry products, thereby, bridging the gap between the demand and supply of products in the market and at the same time serving as a source of income for those who engage in its production (Afolabi, 2002).

*Corresponding Author: Oluwatuyi Toyin Bukola, Department of Agricultural Economics and Extension services, Faculty of Agriculture, University of Benin, Benin City, Nigeria. E-mail: toyintuyi93@gmail.com.

Co-Author Email: [2]omoyore@yahoo.com; [3]osayi.emokpae@uniben.edu

Economic analysis can be defined as the analysis done using payments and value of all items at their value in use or the opportunity cost to the society. It is the comparison of cost with the benefit of a project in order to determine which alternative project has acceptable return, Olayide and Heady, 1982).

Olukosi *et al.* (2007) defined marketing costs as the actual expense incurred in the performance of the marketing functions as commodity moves from the farm to the ultimate consumers. It includes the cost of transport, handling, marketing charges, and cost of assembling, processing, distributions, packaging sales promotion and advertisement cost and other costs such as taxes, levies and excise duties.

Ekunwe and Alufohai (2009) reported in their study that the marketing of egg in Benin City, Edo State, Nigeria was highly profitable and viable; however, their study did not take into consideration the effects of the marketing costs on the magnitude of the marketing margin. This study was conducted to fill in this gap. The specific objectives of this study were to determine the socioeconomic characteristics of the egg marketers, to examine the market conduct and performance of egg marketing, to determine the profitability and marketing margin of egg marketing in the study area, and to determine the effect of marketing cost on the marketing margin.

METHODOLOGY

Study Area

This study was conducted in Oredo and Egor, local government areas of Edo state, Nigeria. This study was carried out in some of the major markets of the local government area where poultry marketing was majorly concentrated. Edo state is an island state in central southern Nigeria, with Benin City as its capital. It is bounded in the north and east by Kogi state, in the south by Delta state, and in the west by Ondo state. Edo state is noted for the following agricultural products; rubber, cashew nuts, cocoa and is blessed with precious stones like quartz, mica, dolomite, granite and lime stone used in cement production.

Sampling Procedures, Data Collection and Data Analysis

A two stage sampling technique involving random and purposive selection was used to select 100 respondents. The first stage involved the purposive selection of two markets from each of the local governments, making it a total of four markets. The second stage involved the random selection of 25 marketers from each market, making a total of 100 marketers. Primary data was obtained using a structured questionnaire and personal oral interview to retrieve information from the respondents.

Data collected were analyzed using descriptive statistics, budgetary techniques and the ordinary least square regression model.

The specific objectives were to; determine the socio economic characteristic of the respondents, determine the conduct and performance of the egg market, estimate the market margin and profitability of egg sellers, determine the effect of marketing cost on the magnitude of the marketing margin.

Descriptive statistics, such as frequency distribution, percentages and mean were used to analyze the socioeconomic characteristics; market conduct was analyzed using descriptive analysis of the egg marketers. Budgetary techniques were used to analyze the market efficiency and profitability of poultry egg marketing.

Profitability Analysis

Profitability analysis was estimated using gross margin analysis, net return, and benefit cost ratio.

Gross Margin = Total Revenue – Total Variable Cost ... (1)

Where: Total Revenue = Price per crates x Quantity of crates sold *and*
Total Variable Cost = Total Cost – Total Fixed Cost

Net return was analyzed using the total profit and is given as
Net returns= Gross Margin – Total Fixed cost ..... (2)

Return per naira was used in assessing the viability of the business and is given as
Return per Naira Invested = Profit/Total Costs invested .(3)

Benefit cost ratio was also used in analysing the viability of the business and the decision rules were
BCR > 1 – The business is viable
BCR = 1 – Break – Even point
BCR<1 – The business is not viable

Benefit Cost Ratio (BCR)
$$= \frac{\text{Present value of total benefit in naira}}{\text{Present value of total costs in naira}} \dots (4)$$

Market Performance Analysis

Market performance was estimated using marketing margin, market efficiency and gross margin analysis, as shown in Equations....

$$\text{Marketing efficiency} = \frac{(\text{Net Profit})}{\text{Total Marketing Cost}} \times \frac{100}{1} \dots (5)$$

Gross Margin = Total Revenue – Total Variable Cost ....(6)
Where, Total Revenue = Price per crates x Quantity of crates sold andTotal Variable Cost = Total Cost – Total Fixed Cost

Marketing Margin = C.P – P.P .... (7)

Where, C.P is consumer price andP.P is producer's price

Ordinary least Square Regression Model

The ordinary least square regression model was used to ascertain the effect of marketing cost on the magnitude of the marketing margin. The implicit model employed was:
$Y = f(X_1, X_2, X_3, X_4, X_5, X_6......e)$ .... (8)

While the explicit form of the equation is
$Y = b_0 + b_1x_1 + b_2x_2 + b_3x_3 + b_4x_4 + b_5x_5 + e$ .... (9)

Where, Y = Marketing margin, X_1 = Transportation cost (₦/month/trader), X_2 = Storage cost (₦/month/trader), X_3 = Market Charges (₦/month/trader), X_4 = Tax/Ticket (₦/month/trader), X_5 = Security Costs (₦/month/trader), e = Error term.

The various constraints affecting egg marketing in the study area was determined using a 5 point Likert scale. The constraints were ranked based on the level of severity into very serious=5, serious=4, moderately serious =3, least serious =2, not serious =1. For a given constraint, the mean was computed by summing the scores of each item and then dividing by the total number of responses. The mean Likert scale point served as a benchmark to know the level of severity of a particular constraint. If the mean was less than three it means that the particular constraint was not serious, while any mean equal to or greater than 3 means that the particular constraint was serious.

RESULTS AND DISCUSSION

Socioeconomic Characteristics

The result in Table 1 indicated that majority of the egg marketers were female, very young and in their active age , married showing that they have responsibilities and households to cater for, this agree with the findings of Ekunwe and Alufohai (2009) who observed that poultry egg marketing in Benin City (Edo, Nigeria)) shows a market dominated by female marketers (96.7%) who have a mean age of 45 years and are married (82.1%). The result also showed that 42.6% had obtained at least secondary school education and also had an average marketing experience of 5 years. This is in consonance with the findings of Mohammed et al. (2013) who found out a mean marketing experience of 5 years among egg marketers in Kuje area of Abuja, Nigeria.

Market Conduct and Performance of Egg Marketing in the Study Area

Market Conduct

Through interactions with the egg marketers in the study area it was found that most of the marketers attracted the consumers through open display and persuasive efforts.

The marketers determined their price by the haggling ability of the consumers, purchase price and quantity demanded. There was, however, no market association for the product so the marketers acted individually.

Market Performance

From the Marketing margin per crates of egg and Market efficiency as seen in Table 2 the result indicated a fairly good market performance.

Market Margin and Profitability of Egg Marketing in the Study Area

Marketing Margin

The result of the average marketing margin as presented in Table 2 shows that there was a price difference of ₦92.62 per crate of egg. Mark up accounted for 70.15% of the total marketing margin while storage cost accounted for 9.07% of the total margin and transportation costs accounted for 7.81% of the total margin.

This shows that the marketers were exploitative in their price as the mark-up took a greater share of the total margin. This finding is in agreement with Ekunwe and Alufohai (2009) report who realized a marketing margin of N60.67($0.17.) and noted that it also took a greater portion of the total marketing margin in Benin, Nigeria.

Profitability analysis of egg marketing in the study area

The result of the gross margin analysis presented in Table 3 shows that on the average, the respondents traded 294 crates per month and this cost them ₦168,162.16 ($460) which represents about 96.38% of their total variable cost. An average egg marketer incurred a total variable cost of ₦174, 478.61($477.28) per month, total revenue of ₦195,393.576 ($534.49). The gross margin of an average egg marketer was ₦20,914.966($57.21) for 294 crates per month while net return per seller was ₦20,622.966 ($56.41) per month. The findings from this study is in agreement with the findings of Okpeye and Ellah (2017) who reported a gross profit margin of ₦772,200($2112.31) and net profit margin of ₦747,500 ($2044.75) per marketer per annum in the South part of Nigeria. The result also shows a return per naira of 0.12 which means that for every N1 invested into the business it yielded N0.12. The result of the marketing margin and profitability analysis therefore shows that egg marketing is a profitable business and should be one to be ventured into.

Viability test of egg marketing

The analysis showed that the business had a BCR of 1.12. The BCR indicated that egg marketing is viable in the study area. This also collaborates with Ekunwe and Alufohai (2009) finding, where they realized that egg marketing in Benin city is very viable (BCR = 1.10).

Effect of marketing cost on the marketing margin

The result of the F value (27.793) was significant (P < 0.05), indicating the goodness of fit of the regression. The result of the regression analysis used in determining the effect of the marketing cost on the magnitude margin is shown in Table 4, the regression estimates shows a fairly good fit for the model as shown by the coefficient of determination (R^2 = 0.597). The adjusted R^2 implies that, the independent variables (transportation cost, storage cost, security cost, and tax and market levy) used in the model, explained about 59.7% of the variations in the marketing margin. The result also showed that transportation costs (X_1) and storage costs (X_2) had a positive effect on marketing margin, which means that the increase in transportation and storage cost will cause an increase in marketing margin while tax had a negative effect, implying that an increase in tax would lead to a decrease in marketing margin and vice-versa. These findings are in agreement with Ekunwe and Emokaro (2009) who reported that transportation cost and other variable were the main component of income realizable from snail marketing in Benin.

TABLES

Table 1: Socioeconomic characteristics of egg marketers

Frequency	Percentage	
AGE		
Less than 35	7	6.9
36-40	12	11.9
41-45	19	18.8
46-50	35	34.7
51-55	24	23.8
56 and above	3	5
SEX		
Female	93	92.1
Male	7	6.9
MARITAL STATUS		
Married	82	82.2
Single	6	5.9
Divorced	5	5
Widow	7	6.9
EDUCATIONAL LEVEL		
Non Formal Education	29	28.7
Primary School	18	17.8
Secondary School	43	42.6
Tertiary Education	11	10.9
EXPERIENCE		
Less Than 5	55	55.4
10-Jun	38	37.7
11 and Above	7	6.9

Marketing constraints faced by the egg marketers

Nine items were ranked on a five Likert type scale in order to estimate the relevant variables that acted as constraints on marketing of poultry eggs in the study area as shown in Table 5. According to the marketers, Broken eggs was the major problem affecting them, and this problem is associated with other problems, such as Inadequate and High Transportation, which was ranked the second most important problem. Poor credit facilities were also identified as a constraint, as well as too many market charges. Theft, High Tax, Lack of market information, Inadequate Storage were of least importance. The finding is in consonance with the findings of Okpeye and Ellah (2017) who found inadequate capital as a great constraint to the marketing of poultry egg in the South part of Nigeria, and Emokaro et al. (2016), who found finance to be one of the major constraints affecting poultry egg production in Benin city, Nigeria. This shows that credit facilities are hard to obtain, discouraging poultry egg marketers and farmers from partaking in the profitable enterprise.

Table 2: Marketing Margin Analysis for the egg marketers per crates in the study area

Items	Amounts (N)
Average transportation costs per crate	7.81
Average security per crate of egg	1.39
Average storage cost per crate of egg	9.07
Average market levy	1.00
Average tax per crate	2.44
Mark-up per crate	70.15
Marketing margin per crate	92.62

Source: Field Survey, 2015

Table 3: Profitability analysis of egg marketers in the study area

Items	Amount (N)	Percentage
Cost of Purchase	168,162.12	96.38
Cost of Transportation	2297	1.32
Storage cost	2668	1.53
Security Cost	346.54	0.20
Market Charges	287.33	0.17
Tax	717.62	0.40
Total Variable Cost	174,478.61	
Fixed cost	292	
Total cost	174,770.61	
Total revenue	195,393.576	
Gross margin	20,914.966	
Net return	20,622.966	
Return per naira	0.12	
Benefit cost ratio	1.12	

Source: Field Survey, 2015

Table 4: Effect of Marketing Cost on the Magnitude of Marketing Margin

Independent variable	Coefficients	Standard error	t-values
Constant	1.513*	0.522	2.900
Transportation	1.055*	0.077	2.536
Storage	1.110*	0.147	2.074
Security	0.222	0.235	0.943
Tax	-0.011	0.005	-0.255
Market Levy	0.038	0.240	0.075

R^2=0.597; F value= 27.793, Source: Field Survey, 2015
* Represents coefficients at 5% significant

Table 5: Constraints faced by the Respondents in The Study Area

Constraints	Mean Value
Theft	2.17
High Tax	1.8
High Transport	3.25*
Poor Credit Facilities	3.50*
Lack of Market Info	1
Too Many Charges	3.09*
Inadequate Storage	2.22
Inadequate Transportation	3.35*
Losses Due to Cracked Eggs	4.19* 4.19 4.19

(Mean value of > 3.00

CONCLUSIONS

The findings from this study showed that poultry egg marketing is dominated by women who are fairly educated and experienced in this field. The results also showed that poultry egg marketing is a profitable business in the study area irrespective of the problems militating against it. It was also shown that too many market charges, poor credit facilities and bad roads, which caused loss of eggs due to cracking, were the problems most faced by the egg marketers. Transportation cost and Storage costs had a positive effect on the marketing margin. Marketing of egg is highly profitable and can also be increased if the problem militating against it is looked into.

RECOMMENDATIONS

Based on the findings of this study, it is highly recommended that governmental and non-governmental organizations set up credit facilities for the people engaged in marketing of poultry eggs. Also, when these credit facilities are available, awareness should be created using agricultural extensions. The egg marketers should be encouraged to form association or cooperative societies, thereby, easing one of the major constraints faced, which is lack of credit facilities, as they can collectively obtain loans at very low interest rates to finance their business.

The government can do their own part by renovating existing bad roads so as to reduce the major constraints faced by the egg marketers and this will help in getting the eggs to the market in time and in good shape.

REFERENCES

Abanikannda OTF, Olutogun O, Leigh AO and Ajayi LA, (2007). Statistical Modelling of Egg Weight and Egg Dimensions in Commercial Layers. Inter. J. Poultry Sci. 6: 59-63.

Afolabi JA (2002). An Analysis of Poultry Egg Marketing in Ondo State Nigeria, Proceeding of the 29th Annual Conference of the Nigeria Society for Animal Production (NSAP), Held at the Federal University of Technology, Akure, March 17th – 21st.

Banerjee, G.C., (1992). Poultry Production, Oxford: Oxford and IBH Publishing Co. PVT. Ltd.

Barbut S (2001). Poultry Product Processing: An industry guide. Boca Raton, Florida: CRC press LLC.

David West KB (2008). Review of livestock development in Nigeria in the 20th century. In Osinowo, O.A., (Ed.) Animal Agriculture in West Africa: Proceedings of the joined silver anniversary conference of the Nigeria Society for Animal production and West African Society for Animal production, inaugural conference, Abeokuta, Nigeria. (pp.127): Abeokuta, Nigeria :Sophie Academic Services.

Ekunwe PA, Alufohai, GO (2009). Economic of Poultry Egg Marketing in Benin City, Edo State, Nigeria. Inter. J. Poultry Sci. 8: 166-169.

Emokaro CO, Akinrinmola FK, Emokpae OP. (2016). Economics of Backyard Poultry Farming in Benin City, Edo State, Nigeria. Nigeria.J.Agric.Food and Envr . 12:50-57.

Emokaro CO, Ekunwe, PA. (2009). Determinants of Income Realizable from Snail Marketing in Benin City, Nigeria. J. Agric. Forestry and Fisheries. 10: 35-38.

Esingmer, M.E., 1991. Animal Science Series, Daurille, Illnois: Interstate Publishers.

Mohammed AB, Mohammed SA, Ayanlare AF, Afolabi OK. (2013). Evaluation of Poultry Egg Marketing in Kuje Area Council Municipality of F.C.T Abuja, Nigeria. Greener J. Agric Sci. 3: 068-072

Oji UO, and Chukwuma AA. 2007. Technical Efficiency of Small Scale Poultry-Egg Production in Nigeria: Empirical Study of Poultry Farmers in Imo State, Nigeria. Research J. of Poultry Sciences, 1: 16-22.

Okpeye MY, Ellah GO. (2017). Analysis of Poultry Egg Marketing in South South Part of Nigeria. Global J. of Agric.Research. 6: 1-15.

Olayide SO, Heady EO (1982) Introduction to Agricultural Production Economics,University Press Ibadan pp 17-18.

Olukosi JO, Isitor SU, Ode MO (2007). Introduction to Agricultural Marketing and Prices. 3rd Edition. Principles and Application.

5

Maize seed marketing chains and marketing efficiency along supply chains of the hills in Nepal

K.C. Dilli Bahadur[1*], N. Gadal[2], S. P. Neupane[3], R.R. Puri[4], B. Khatiwada[5], G. Ortiz-Ferrara[6], A.R. Sadananda[7], C. Böber[8]

[1*, 2, 3, 4, 5, 6, 7, 8]International Maize and Wheat Improvement Center (CIMMYT Intl., Mexico), CGIAR Center, Nepal.

Remoteness, poor infrastructures, labor shortages, small quantities of seed at the producer level and few private seed traders are inherent problems in maize seed production and marketing in the hills of Nepal. Farm-saved seed, including seed exchange and private sector supply are the main sources of improved maize seeds in Nepal. Using the primary data collected from 200 respondents across 20 hilly districts of Nepal, this paper analyzes marketing chains and the efficiency of marketing of improved maize seed along the supply chains.

The results show five major maize seed marketing chains. Chain I involved producers, collectors, wholesalers, retailers and consumers; Chain II involved producers, collectors, wholesalers and consumers; Chain III involved producers, collectors, retailers and consumers; Chain IV involved producers, collectors and consumers; and Chain V involved producers and consumers. A total of 64.3 tons of improved maize seed was marketed through the identified chains. Chain II was the most important supply chain, accounting for 38.8per cent of total marketed seeds; while Chain I was the least important, accounting for 4.3per cent. Producers' share on consumer price was highest in Chain V (100per cent) and lowest in Chain III (66per cent). Transportation cost accounted for the highest amount (average 47.5per cent). Highest margin of profit (NRs 6.5/kg) was taken by retailers and lowest by collectors (NRs 2.5/kg). Highest marketing efficiency with a magnitude of 7.24 was observed in Chain V and lowest with a magnitude of 0.9 in Chain I.

Keywords: Community-based seed production, agrovets, marketing chain, marketing efficiency, seed traders.

INTRODUCTION

Maize (*Zea mays L.*) is the second most important cereal crop in Nepal in terms of area and production. It is grown on about 0.85 million hectares and production is 1.99 million tons (MoAD, 2013). About 78per cent of the total maize cultivated area is in the hills. Maize is a major cereal crop in the mid-hills of Nepal, particularly among poorer families and disadvantaged groups. In the hills, maize is the main source of livelihoods – food, feed, fodder and fuel. And there is a common saying "*if there is no maize... there is nothing to eat*". Maize yield in Nepal is 2.35 tons per hectare (t/ha) (MoAD, 2013) against the attainable yield of 5.7 t/ha (Gurung, 2012). Global maize productivity is 5.52 t/ha and in neighboring countries, India's average is 2.45 t/ha and the average in Bangladesh is 7.0 t/ha (FAO, 2013). The productivity of maize in Nepal is constrained primarily by poor access to improved varieties of seed, fertilizer, shortage of labour and farmers' lack of awareness about new maize production technologies, etc.

Corresponding Author: KC Dilli, International Maize and Wheat Improvement Center (CIMMYT Intl., Mexico), CGIAR Center. Email: d.kc@cgiar.org

Despite the past 50 years of seed development initiatives, unavailability of good quality seeds in sufficient amount at right time and place is the main constraint for improving lives of many Nepali farmers. The formal seed sector in Nepal has been incapable of narrowing the gap between seed requirements and the production of improved maize seed. Informal seed supply system provide more than 90per cent planting materials in the country (PACT, 2012). Farm-saved seed, including seed exchange, and private sector supply are the main sources of improved maize seeds in Nepal (KC, 2013). The majority of the improved maize seed producers in Nepal are located in very remote areas and small enterprises. Inherent problems in seed production and marketing at the farmer level are the small marketable surplus of seed at the individual level, poor accessibility due to remoteness and high transportation costs; these make the sector less attractive to private seed traders. The limited presence of private seed traders (agrovets and seed companies) results in very weak seed markets, particularly in the hills.

The overall goal of this paper is to assess seed marketing chains and analyze the marketing efficiency of improved maize seed produced under a community-based seed production (CBSP) system in the hills of Nepal. Meeting the overall goal, price spread, marketing costs and margin along the seed supply chains were examined.

In Nepal, seed multiplication of improved maize varieties through the Community-Based Seed Production (CBSP) program was initiated in the Hill Maize Research Project (HMRP) in 2000. This has been a very successful model, which significantly contributed to increase the production of improved seeds and has increased the seed replacement rate. In 2000, seven CBSP groups produced 14 tons of improved maize seed and in 2013 a total of 1,216 tons of improved seed was produced through 223 CBSP groups (HMRP, 2013). During this period the annual compound growth rate of maize seed under the CBSP program was 34.5per cent (KC, 2013). The maize seed replacement rate in Nepal increased from 5.81per cent in 2007 to 11.3per cent in 2011 (Pokharel and SQCC, 2012).

CBSP follows the approach of forming a community-based farmer group comprising 15 to 25 members and registered in the District Agriculture Development Office (DADO). The CBSP group produces improved maize seed of most preferred varieties best suited to the locality. This approach also aimed to strengthen the capacity of the local community as seed producers to primarily undertake seed production and marketing activities through the maturation of these groups into cooperatives and then to establish private seed companies for seed trading. The ultimate target of this approach is to increase the role of the private sector in maize seed production and marketing in a sustainable way.

MATERIALS AND METHODS

In general, the study methodology involve assessing the structure of the maize seed marketing chains specifically exploring: what are the marketing chains?; who are the chain actors?; what are their functions?; and so on. More importantly it involved identifying the market chain within it–all who contributed to seed assembling, processing, grading, packaging, labeling and marketing of maize seed.

Data source

The study used primary data collected during the period September 2012 to March 2013 from 200 respondents including 178 CBSP farmers and 22 seed traders (cooperatives, seed banks/seed companies, agrovets). A purposive random sampling technique was used to select the sampling units. Twenty HMRP districts, where improved maize seed is produced in the communities, were selected purposively for the study. Data were collected through sample survey method. Focus group discussions (FGDs) and a key informant survey (KIS) were also conducted with related stakeholders to collect quantitative and qualitative information on several aspects of seed marketing.

Analytical methods

Marketing chain

A marketing chain is the chain of intermediaries through whom the product passes from producers to the ultimate consumers. Kohls and Uhl have defined marketing chains as alternative routes of product flows from producers to consumers.

The assessment of maize seed marketing chains was accomplished by analyzing the seed flow through various chains. Seed marketing analysis covered the services involved in assembling seed from the production point to the ultimate seed user (i.e. the consumer). The quantity of seed disposed through identified chains was comprehensively analyzed. In the marketing chain numerous interconnected activities like assembling, transport, processing, grading, packaging, labeling, storage and sale were assessed along the market chains.

Price spread

Price spread refers to the difference between the price paid by the consumer and price received by the producer. Price spread consists of marketing costs and margins of the intermediaries (Prasad, 1989). Added costs and margins of the market intermediaries were calculated as the per cent of the buyer's price along the chain. The producer's share in consumer's price was estimated as per Prasad (1989) and Joel et al. (2013).

Marketing cost and margin

The major costs involved in the maize seed marketing chain were seed handling cost or load/unload, transportation, processing, grading, packaging, labeling, storage, etc. Marketing cost at various market actors was analyzed by taking the average cost of each activity incurred by market actors and expressed in NRs/kg. Marketing margins can also be categorized into gross and net margin. The gross marketing margin at the trader level can be explained as the difference between the sale price and the purchase price of that trader or the sum of incurred marketing costs and margin of profit taken by that trader. The gross marketing margin consists of the costs in moving the produce from the point of production to the consumers, whereas net marketing margin is simply the margin of profit of the trader taken against operating the business and expressed in NRs/kg.

Marketing efficiency

Marketing efficiency is the ratio of market output (satisfaction) to marketing input (cost of resources). An increase in the ratio represents improved efficiency and vice versa. In other words, it is the price received by the producer and expressed as a percentage of the consumer price. The efficiency of a marketing system is measured in terms of costs to the system of inputs to achieve a given amount of output. The Shepherd's Index formula (1965) was employed to estimate the marketing efficiency. The higher the ratio implies a higher marketing efficiency, or the chain is said to be more efficient and vice versa.

RESULTS AND DISCUSSIONS

Marketing Functions

Seed collection. Seed produced by the CBSP groups was assembled by seed collectors. A total of 64.3 tons of seed was entered into the marketing chain. Of this 53.4 tons (83per cent) of seed was collected by collectors (mainly cooperatives) and the remaining quantity was either directly sold or exchanged with maize-cultivating farmers (consumers) through the seed producers. Manakamana 3, Deuti, Rampur Composite, Manakamana 4, Arun 2, Manakamana 5 and Poshilo Makai 1 (QPM) were the common varieties of seed trading in the identified market chains. Seed from producers up to the collection point was transported primarily by porters, whereas transport from collector up-to retailer was done by pick-ups, small trucks and buses.

Processing and grading. Processing was carried out mainly to separate inert materials from the seed lot, whereas grading was done to sort seed into different lots, each of which with substantially uniform quality characteristics. The presence of a grading system makes possible a comparison of values of different qualities of a product in a single market and of differences in price of the same grade in different markets. After collecting the seed from the producer it was processed and graded (mostly manually) at the collector level. Grading was done mainly on the basis of size and color of seed. Sunken and broken seeds were removed during processing.

Packaging and labeling. Packaging and labeling are very important functions in seed marketing. Packaging provides product safety and offers a presentable form to consumers; labeling provides required information to the consumers about the produce. While collecting seed from producers up to the collectors' point, seed was packed into ordinary gunny bags of varying sizes ranging from 25 to 50 kilograms (kg). Packaging at the wholesaler level was also done in gunny bags and at the retailer level mostly in cotton cloth bags (1 to 10 kg). Before marketing, each sac/bag of seed was truthfully labeled, complying with the conditions set by the Seed Act and Seed Regulation of Nepal. It was noticed that truthful labeling was made mandatory to ensure tracking of producer or seller, in case of any malpractices.

Seed storage. Proper storage of seed at good facilities is an important feature of efficient marketing. To maintain longer viability, seed should be stored in a well-ventilated and damp-proof room. In general, due to lack of proper storage facilities, seed producer farmers usually sell seed immediately after harvest at lower rates; however, some farmers store seed in metal bins or in gunny bags until traders come. In regard to the sample traders, seed at the collector level was stored in metal seed bins; whereas at the wholesalers' level it was stored in large size gunny bags (40-50 kg) and at the retailer level some seed is stored in small drums and some in 1-10 kg cloth or plastic bags.

Seed trading. Three types of seed traders were found in the study area – collectors, wholesalers and retailers. It was inferred that the collectors (mostly cooperatives) collected seed from producers; community seed banks/private seed companies performed wholesaling; and agrovets and some groceries retail seed at the local level. It was observed that the collectors assembled seed (53.4 tons) from seed producers. After processing and packaging, seed was sold to wholesalers (27.8 tons), retailers (9.6 tons) and directly to seed users/consumers (16.0 tons), either purchased or distributed by government organizations such as District Agriculture Development Office (DADOs), Village Development

Figure 1. Maize seed marketing chains and flow of seed volume in the study area (2012-2013).

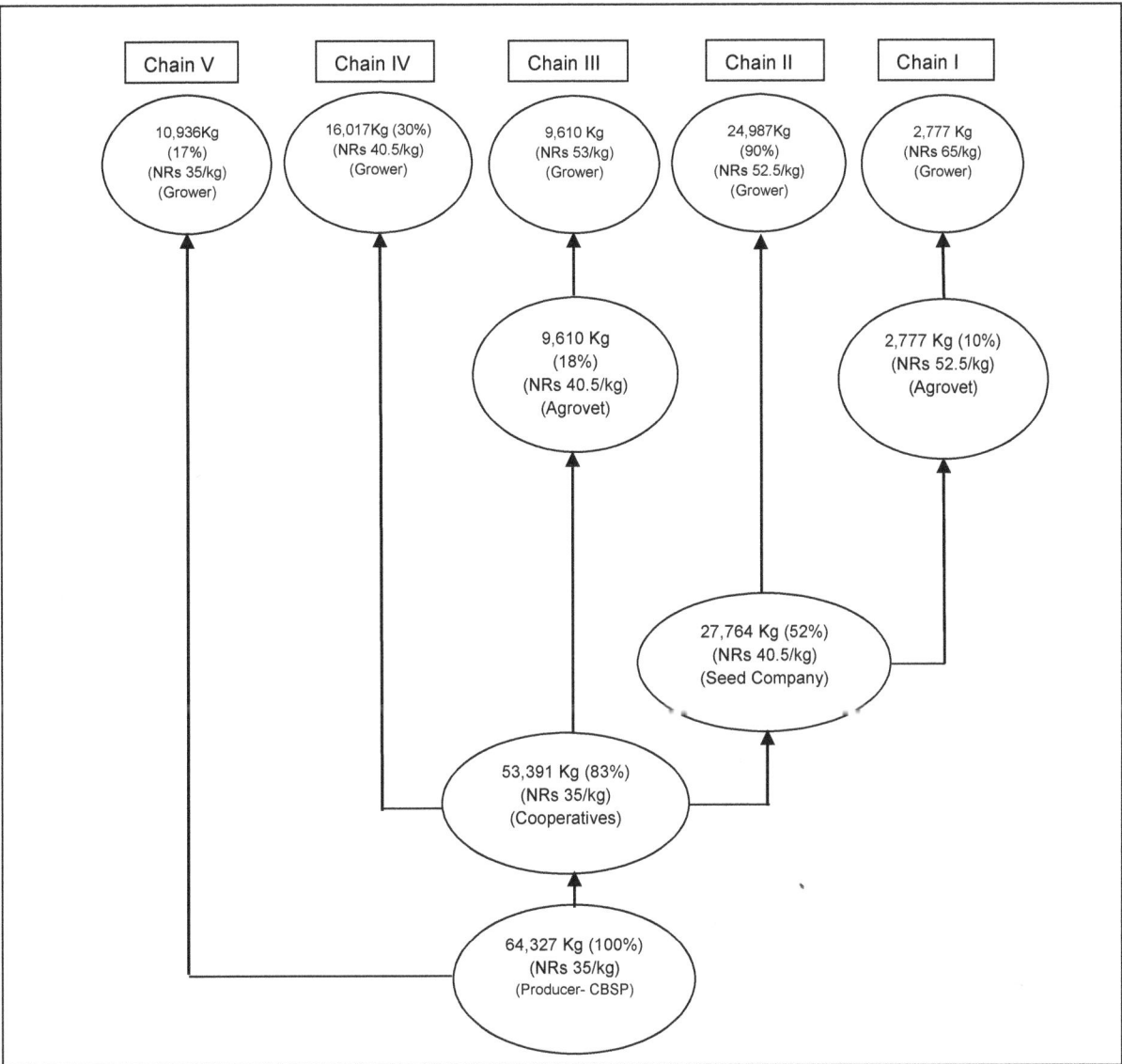

Exchange rate: 1 US$=Nepalese Rupees (NRs) 95.0

Committee (VDCs) (GOs) and international/national non-governmental organizations (I/NGOs). Of the total seed traded by the wholesalers, about 25.0 tons was directly distributed to maize grower farmers through GOs and I/NGOs and the remaining 2.8 tons was purchased by the maize grower farmers through agrovets (Figure 1).

Marketing Chains

The results show five major maize seed marketing chains. Chain I involved producers, collectors, wholesalers, retailers and consumers; Chain II involved producers, collectors, wholesalers and consumers; Chain III involved producers, collectors, retailers and consumers; Chain IV involved producers, collectors and consumers; and Chain V involved producers and

consumers. A total of 64.3 tons of improved maize seed was marketed through these market actors. It was observed that of the total traded quantity of seed, about 83per cent was sold/distributed through formal chains (cooperatives, seed companies, agrovets). Trading of maize seed through identified market chains is presented in Table 1.

It may be observed from Table 1 that Chain II was the most important supply chain, accounting for 38.8per cent (25 tons) of total marketed seeds and Chain I was least important, accounting for only 4.3per cent (2.8 tons).Likewise, Chain IV accounted for 24.9per cent, Chain V accounted for 17.0per cent and Chain III accounted for 15.0per cent of the total marketed seed volume.

Table 1. Disposal of improved maize seed through identified Chains

Chain	Marketing chain actor	Quantity of seed flow (kg)	% of total quantity
I	Producer to Collector to Wholesaler to Retailer to Consumer	2,777	4.3
II	Producer to Collector to Wholesaler to Consumer	24,987	38.8
III	Producer to Collector to Retailer to Consumer	9,610	15.0
IV	Producer to Collector to Consumer	16,017	24.9
V	Producer to Consumer	10,936	17.0
	Total	64,327	100.0

Table 2: Price received by producer, marketing cost, margin of profit, price paid by consumer in the identified market chains (Amount in NRs/kg)

Market intermediaries	Chain I		Chain II		Chain III		Chain IV		Chain V	
	Amount	% Share	Amount	% Share	Amount	% Share	Amount	% Share	Amount	% Share
Price received by producer	35.00	53.8	35.00	66.7	35.00	66.0	35.00	86.4	35.00	100.0
Total marketing cost	17.50	26.9	11.50	21.9	9.00	17.0	3.00	7.4	-	0.0
Total margin of profit	12.50	19.2	6.00	11.4	9.00	17.0	2.50	6.2	-	0.0
Price paid by consumer	65.00	100.0	52.50	100.0	53.00	100.0	40.50	100.0	35.00	100.0

The larger proportion of seed sold through Chain II was probably due to large quantities of seed purchased and distributed by GOs and I/NGOs to maize grower farmers under their program such as conducting result and method demonstrations and a seed for food security program.

The GOs and I/NGOs normally purchase seed just before sowing time from the seed companies when there is no seed available at the farmer level but still available at these seed companies. Another reason was GOs and I/NGOs prefer to purchase seed from formal institutions such as cooperatives and seed companies rather than from CBSP groups and agrovets to maintain transparency and to collect large quantities of seed from one place, even if the price is relatively higher than the other organizations. A comparatively lower proportion of sales through Chain I (4.3per cent) was due to the fact that the majority of farmers have very limited information about agrovets selling maize seed. More importantly, the presence of agrovets is very limited in the hills and they prefer to sell hybrid seed rather than open pollinated varieties because they fetch higher profits with hybrids.

Price Spread

The price spread comprises all details of various marketing costs as well as the margin of profit taken by market intermediaries present in the marketing chains at subsequent stages. The price spread in the marketing of maize seed in the study area was worked out and presented in Table 2.

Average price per kg received by producers in all chains

was NRs 35 (Table 2). On average, the price paid by consumers was highest in Chain I (NRs 65/kg) and lowest in Chain V (NRs 35/kg). The price per kg paid by consumers in chains II, III and IV was NRs 52.5, 53.0 and 40.5, respectively. The producers' share in consumers' rupee varied from 53.8per cent to 100per cent. Cent per cent of producers' share in consumers' rupee in Chain V was mainly due to the fact there were no market intermediaries involved in this chain, seed was directly purchased by or reached to consumers from producers; therefore, no marketing costs and margins were incurred to the consumers' price. This finding is similar to Joel et al.(2013) where they observed highest producer share to consumer price (94 per cent) in similar maize grain chain "producers to final consumers" and reported due to the fact that market intermediaries who pocketed the producer share were eliminated along the maize grains supply chain. The lowest price spread was observed in Chain I, where the producer received 53.8percent of the amount paid by the consumer (NRs 35 out of NRs 65/kg). This was due to higher marketing costs (26.9per cent) and margins (19.2per cent) incurred by market intermediaries. Similarly, the producers' share in consumer rupees in chains II, III and IV was 66.7per cent, 66.0per cent and 86.4per cent, respectively (Table 2).

Per kg seed marketing costs varied from NRs 3.0 (Chain IV) to 17.5/kg (Chain I), accounting for 7.4 per cent and 26.9 per cent of consumers' prices in respective chains (Table 2). Per kg marketing margin taken by intermediaries varied from NRs 2.5 (Chain IV) to NRs 12.5/kg (Chain I), which accounted for 6.2per cent and 19.2per cent of the price paid by the consumer in respective chains. It is also worth pointing out that

Table 3. Marketing costs at market intermediaries in the identified chains (Amount in NRs/kg)

Market intermediaries	Chain I		Chain II		Chain III		Chain IV		Chain V	
	Amount	% Share	Amount	% Share	Amount	% Share	Amount	% Share	Amount	% Share
Collector	3.00	17.1	3.00	26.1	3.00	33.3	3.00	100.0	NA	NA
Wholesaler	8.50	48.6	8.50	73.9	-	-	-	-	NA	NA
Retailer	6.00	34.3	-	0.0	6.00	66.7	-	-	NA	NA
Total added costs	**17.50**	**100.0**	**11.50**	**100.0**	**9.00**	**100.0**	**3.00**	**100.0**		

Table 4. Marketing costs on various marketing functions in the identified chains (Amount in NRs/kg)

Cost item	Chain I		Chain II		Chain III		Chain IV		Chain V	
	Amount	% Share	Amount	% Share	Amount	% Share	Amount	% Share	Amount	% Share
Load/ unload charge (handling)	1.15	6.6	0.65	5.7	0.75	8.3	0.25	8.3	NA	NA
Transportation charges	8.10	46.3	5.35	46.5	4.25	47.2	1.50	50.0	NA	NA
Processing and grading carges	1.25	7.1	0.75	6.5	0.75	8.3	0.25	8.3	NA	NA
Packaging including material	4.50	25.7	3.25	28.3	2.00	22.2	0.75	25.0	NA	NA
Storage charges	2.50	14.3	1.50	13.0	1.25	13.9	0.25	8.3	NA	NA
Total added costs	**17.50**	**100.0**	**11.50**	**100.0**	**9.00**	**100.0**	**3.00**	**100.0**		

retailers sold the seed at a comparatively higher rate that is NRs 65/kg compared to wholesaler (NRs 52.5/kg) and collector (NRs 40.5/kg). It was mainly due to the fact that the retailers sold the seed in small quantities and were required to retain the seed for a longer period of time which involves incurring higher risk.

Marketing Cost

It may be observed form the analysis that no marketing cost and margin was incurred in Chain V because seed was directly purchased by or reached to consumers from seed producers; no market intermediaries were involved in this chain (Table 3). Per kg marketing cost incurred by various market intermediaries was analyzed at NRs 17.5, 11.5, 9.0 and 3.0 in chains I, II, III and IV, respectively. It was observed that wholesalers' marketing cost was relatively high (NRs 8.5/kg), which accounted for 48.6per cent and 73.9per cent of the total costs in Chain I and II, respectively. The lowest marketing cost was observed at the collector level (NRs 3.0/kg), which accounted for 17.1per cent, 26.1per cent, 33.3per cent and 100per cent of total cost in chains I, II, III and IV, respectively. Retailers' marketing cost was NRs 6.0/kg, which accounted for 34.3per cent and 66.7per cent of the total cost in chains I and III, respectively. Higher wholesalers' marketing cost was mainly due to higher transportation costs and seed packaging, including materials (Table 3 and Table 4).

Per kg marketing cost incurred carrying out various marketing functions by market chain is presented in Table 4. An item-wise analysis indicated that the cost of transportation accounted for the highest share – on average 47.5per cent (NRs 1.5 to 8.1/kg) of the total marketing cost in the identified chains. The cost of the load-unload charge accounted for the lowest; on average 7.2per cent (NRs 0.25 to 1.15/kg) of total marketing cost in the identified chains. Other major marketing cost items

were: the cost of packaging (including materials), on average accounted for 25.3per cent (NRs 2.6/kg); storage charges (12.4per cent or NRs 1.4/kg), and processing and grading costs– 7.6per cent or NRs 0.75/kg in identified chains (Table 4).

Marketing Margin

The marketing margin is also an indicator of the efficiency of the marketing system. The larger the value of marketing margins the greater the inefficiency in the marketing system. On the other hand, if the goods move from producers to consumers at minimum cost, the marketing system can be said to be efficient. However, in such a situation the sustainability of the marketing system might remain questionable.

Per kg gross margin was NRs 34.2, 21.7, 22.2, 9.7 and 4.2 in chains I, II, III, IV and V, respectively (Table 5). Retailers' gross margin was highest (NRs 12.5/kg), which accounted for 36.5per cent and 56.2per cent of the total gross margin in chains I and III, respectively. Gross margin by intermediaries showed that producers incurred the lowest margin NRs 4.2/kg (on top of cost of production) accounted for 12.4, 19.5, 19.1, 43.6 and 100per cent of total gross margin in chains I, II, III, IV and V, respectively. Per kg gross margin of seed collectors and wholesalers was calculated to be NRs 5.5 and 12.0, respectively. The higher retailer margin was mainly due to the fact that the disposal of seed at the retail level was slow, as well as in small quantities and storage problems compared to collectors and wholesalers, resulting in comparatively higher risk and loss.

Per kg total net marketing margin (margin of profit) was NRs 16.7, 10.2, 13.2, 6.7 and 4.2 in chains I, II, III, IV and V, respectively (Table 5). As explained below in gross margin, retailers' net margin was relatively high (NRs 6.5/kg), which accounted for 38.8per cent and 49.1per

Table 5. Gross and net margins at various market actors in the identified chains (Amount in NRs/kg)

Market intermediaries	Chain I		Chain II		Chain III		Chain IV		Chain V	
	Amount	% Share	Amount	% Share	Amount	% Share	Amount	% Share	Amount	% Share
Total gross margin (GM)	**34.25**	**100.0**	**21.75**	**100.0**	**22.25**	**100.0**	**9.75**	**100.0**	**4.25**	**100.0**
Producer	4.25	12.4	4.25	19.5	4.25	19.1	4.25	43.6	4.25	100.0
Collectors	5.50	16.1	5.50	25.3	5.50	24.7	5.50	56.4	-	-
Wholesalers	12.00	35.0	12.00	55.2	-	-	-	-	-	-
Retailer	12.50	36.5	-	-	12.50	56.2	-	-	-	-
Total net margin (margin of profit)	**16.75**	**100.0**	**10.25**	**100.0**	**13.25**	**100.0**	**6.75**	**100.0**	**4.25**	**100.0**
Producer	4.25	25.4	4.25	41.5	4.25	32.1	4.25	63.0	4.25	100.0
Collectors	2.50	14.9	2.50	24.4	2.50	18.9	2.50	37.0	-	-
Wholesalers	3.50	20.9	3.50	34.1	-	-	-	-	-	-
Retailer	6.50	38.8	-	-	6.50	49.1	-	-	-	-

Table 6. Index of marketing efficiency in the identified market chains

Item	Chain I	Chain II	Chain III	Chain IV	Chain V
Consumers price (NRs/kg)	65.00	52.50	53.00	40.50	35.00
Total marketing costs and margin of profit (NRs/kg)	34.25	21.75	22.25	9.75	4.25
Marketing Efficiency (Index)	**0.90**	**1.41**	**1.38**	**3.15**	**7.24**

cent of total net margin in chains I and III, respectively. The lowest net margin was charged by collectors (NRs 2.5/kg) accounted for 14.9, 24.4, 18.9 and 37.0per cent of total net margin in chains I, II, III and IV, respectively. Low net margin at the collector level might be due to competition between seed collectors.

Marketing Efficiency

The marketing efficiency is directly related to the cost involved to move produce from the producer to the consumer and required services provided or desired by the ultimate consumers. If the cost compared with the services involved is low, such marketing said to be efficient and vice-versa. More importantly, any improvement that reduces the cost of a particular function without reducing consumers' satisfaction indicates an improvement in the marketing efficiency. Highest marketing efficiency with magnitude of 29.05 in maize grain chain "producers to final consumers" was influenced by elimination of market intermediaries from the chain (Joel et al., 2013).

The results presented in Table 6 revealed that highest marketing efficiency index (7.24) was in Chain V (producer to consumer), followed by Chain IV (producer to wholesaler to consumer) with a marketing efficiency index of 3.15, followed by Chain II (producer to retailer to consumer) with a marketing efficiency index 1.41, followed by Chain III with a marketing efficiency index of 1.38 and lastly Chain I (producer to wholesaler to retailer to consumer), with a marketing efficiency index of 0.9. The highest marketing efficiency in Chain V was mainly due to an absence of market intermediaries, since

consumers either purchased or received seed directly from producers or suppliers (GOs or I/NGOs). However, it was observed that the higher proportion of maize seed was marketed through Chain II (38.8per cent of total seed transaction). It must be probably due to the fact that there were larger number of CBSP-managed co-operatives (seed collector) in the study area and the consumers were of the opinion that the seed purchased from cooperatives assured quality of seed.

CONCLUSION

Maize seed production through the CBSP approach in Nepal can be considered a means to disseminate improved seed in remote areas. The CBSP groups are helping solve the problem to supply quality seed on time in demanded quantities. So the government must meet the growing need to up-scale and out-scale the CBSP approach to produce quality seed under the agriculture extension service system of the Department of Agriculture, Nepal.

Seed marketing through formal chains (83per cent) is quite satisfactory and the market chain is being developed, although there is potential for even more. Of the identified five seed marketing chains, the cheapest rate of seed (NRs 35/kg) was purchased in Chain V (market efficiency 100per cent). This is obvious since there were no market intermediaries present; seed was directly purchased or obtained by consumers from the producers. However, sustainability of this market chain might be at question in the absence of market intermediaries in such direct transactions. So, previewing the market sustainability at least one market intermediary should be involved between producer and consumer as in Chain IV (producer to collector/cooperative to consumer) where the market efficiency index (3.15) was in second rank. The Government of Nepal (GoN) is also giving high priority to establish cooperatives in remote areas for this type of business.

A higher magnitude of gross marketing margin is not only an additional burden on consumers but also an injustice to the producers who do not get a reasonable benefit from their produce and ultimately may have an adverse effect on production.

The largest unit profit (NRs 6.5/kg) which was observed at the retailer level should be reduced by introducing a safety cushion such as an insurance policy. The provision of counterpart insurance premium from government to seed producers on crop failure caused by natural calamities like drought, hailstones, landslides and floods and subsidies to seed traders for transport, tax exemption on processing and grading equipment import, advertisement and provision of financial loan against seed stock could help in lowering the surcharge on seed.

Overall, the maize seed marketing scenario indicated that on average 25.4per cent of the consumers' price was absorbed in the marketing chains in meeting out the various marketing costs (14.6per cent) and margin of profit of traders (10.8per cent), and 74.6per cent of consumers' price reaches in the hands of the seed producer farmers. Of the total amount received by seed producers, about 40.6per cent is spent on seed production to cover variable and fixed costs and 6.8per cent of the amount was spent on marketing costs, leaving only about 9per cent as a margin of profit to cover enterprise risks. The main reason for the comparatively lower price obtained by the seed producers on the one hand and higher prices paid by the consumers on the other hand, is due to the existence of a large number of market intermediaries resulting in a larger magnitude of gross marketing margins. Furthermore, seed producers spent a substantial amount of time and huge resources, as well as taking higher risks to produce seeds. But their share in total profit is small. On the other hand, compared to seed producers, seed collectors, wholesalers and retailers spent relatively less time and invested fewer resources, but they received large profits. In reality, producers are getting a low price, consumers are paying high prices and most benefits go to traders. Profits at the producer level could be increased by providing a counterpart subsidy from the government to cover some portion of marketing costs at the trader level.

Several findings show that fewer actors in the seed marketing chain reduce the retail price. It can benefit both producers and consumers. One way to reduce the number of actors is to directly link producers to consumers. Another way to benefit both producers and consumers is by increasing the number of traders, which will increase market competition and reduce the monopoly power of a few traders. This requires investment in infrastructure, market information, capacity building of seed producing groups, and establishing linkage between producers' groups and seed traders.

According to several empirical findings, the market intermediaries reduce marketing efficiency of seed along the supply chain. Therefore, it is the role of the government to eliminate market intermediaries along the seed supply chain who pocketed a large share of the margins of seed producers. However, an optimum number of market intermediaries is required for the sustainability of the seed marketing system. Seed producers could maximize their margins if government intervenes proactively in order to establish/organize and streamline the seed marketing cooperatives and federations. Producers could then use these federations to sell their seed at better profit through spot and contract seed production.

ACKNOWLEDGEMENT

Great appreciation to Hari P. Sharma, Kiran Basnet, Sabitri Dhakal, Anju Pandey, Kamala Sapkota, Santosh Rasaily, Sunil Chaudhary, Shanti Pandey, Abiskar Gyawali, Prameela Awale and Urmila Adhikarifor their cooperation in conducting the interviews. Immense thanks to Ambika Pandey for her great support in data compilation. Special thanks to Dr. H. N. Bhandari for his critical inputs and valuable suggestions. Sincere thanks to Dr. B. M. Prasanna for his encouragement and continuous support while carrying out this study. Finally, great thanks to survey respondents and HMRP partners for their cooperation accomplishing this study.

REFERENCES

Joel C, Vasudev N, Suhasini K (2013). Marketing efficiency of agri-food along the agri-food supply chain in Tanzania, International Researchers. 2(1):127:131www.iresearcher.org. Accessed June5, 2014.

Dilli KC, Ortiz-Ferrara G, Gadal N, Gurung DB, Pokharel S (2011). Economics of maize seed production, marketing and value chain system under CBSP system in the Hills of Nepal. 11th Asian Maize Conference, CIMMYT Intl., Nanning, China.

Dilli KC (2011). Time series analysis of growth rates of area, production, and productivity of major cereal crops of Nepal. Annual Review Meeting, Hill Maize Research Project, Nepal.

Dilli KC (2013). Maize seed production and marketing: A value chain analysis (in Nepali Language). Our Heritage. 4:9-13.

Dilli KC (2013). Maize seed production communities in hills towards a new path of contract seed production in Nepal. Agronomy Journal of Nepal. 3:151-156.

Dilli KC, Gadal N, Sadananda AR, Boeber C, Koirala KB, Neupane SP, Khatiwada B, Basnet K (2013). Value chain analysis of community based maize seed production in the hills of Nepal. 12th Asian Maize Conference, CIMMYT Intl., Bangkok, Thailand.

MacRobert JF (2009). Seed business management in Africa. Harare, Zimbabwe, International Maize and Wheat Improvement Center (CIMMYT); 2009.

Ministry of Agricultural Development, Nepal. (2013). Statistical Information on Nepalese Agriculture; 2013.

Lilian K, Nicholas S, Jayne TS, Francis K, Milu M, Megan S, James F, Gilbert B (2011). A farm gate-to-consumer value chain analysis of Kenya's maize marketing system. Michigan State University, Working Paper No. 111.

Prasad J (1989). Marketable surplus and market performance: A Study with Special Reference to Muzaffarpur Food-grain Market in Bihar, India. Mittal Publications, New Delhi; 1989.

Project for Agriculture Commercialization and Trade (2012). Value Chain Development Plan for Cereal Seed., MoAD, Nepal; 2012.

6

Analysis of Yam Marketing in Akoko North-East Local Government Area of Ondo State, Nigeria

Rabirou Kassali[1*], Abdulhameed Abana Girei[2] and Ismaila Dauda Sanu[3]

[1]Department of Agricultural Economics, Obafemi Awolowo University, Ile-Ife, Osun State, Nigeria
[2,3]Department of Agricultural Economics and Extension, Nasarawa State University, Keffi, Nigeria

The study analysed the marketing of yam in Akoko North East Local Government Area of Ondo State. It specifically described the socio-economic characteristics of yam marketers and determined the marketing margin, marketing cost, markup, operational efficiency and the constraints faced by yam marketers in the area. Data used for the study were generated through the administration of structured questionnaire. A total of 90 respondents comprising of 30 yam wholesalers and 60 retailers, were randomly sampled from three purposively selected major yam markets. Data were analyzed using descriptive statistics, Concentration Ratio, Gini Coefficient and Operational Efficiency Model. The results showed variation in marketing cost, marketing margin, marketing profit, and markup for both wholesale and retail yam markets. Gini Coefficients of 0.307 and 0.307 were obtained for wholesaler and retailer respectively. This indicated high level of competition in the industry. Retailers were more operationally efficient than wholesalers, but wholesalers' marketing profit and markup were higher than that of retailers. Wholesalers complained of insecurity, price uncertainty, high cost of yam and, transportation cost. Also, high capital requirement is considered as a serious constraint, while retailers complained of high cost of yam, price uncertainty, capital intensity and insecurity as very serious constraints. Alleviating some of the challenges, would therefore lead to more efficiency in yam business, while improving on the welfare of yam marketers.

Keywords: yam, wholesaling, retailing, market structure, marketing performance

INTRODUCTION

Yam (Dioscorea spp) is an herbaceous vine that grows under tropical climates around the world. It is a tuberous root whose skin colour varies from dark brown to light pink (Huxley, 1992) with size varying from an average 2.5 kg to 5 kg (Kay, 1987) according to varieties and growing conditions. It is a starchy tuber rich in carbohydrates, with flesh colour ranging from white to purple, through yellow or pink (Huxley, 1992). According to FAO (2014), the world output of yam is estimated at 68.1 million tones on a total cultivated land area of 7.8 million ha, corresponding to an average yield of 8.8 t/ha. Nigeria is the largest producer of yams with about 45 million tonnes in 2014, representing 66 percent of world output.

According USDA Nutrient Database, Yam is an important source of nutrients. One hundred (100g) gram of yam contain 494kj of energy, 27.9g of carbohydrates and 4.1g of dietary fiber. Yam is also an important source of minerals (Potassium, Calcium, Phosphorous, Magnesium, etc.) and vitamins (Vitamins; C, B complex, folate.). It is characterized by low glycemic index (Atkinson et al., 2008) and constitutes one of the most readily available crop based healthy protein among the poor in tropical areas, most especially in Nigeria. Yam provides about 200 calories of energy per day per capita (Babaleye, 2003). In term of recipes, yam is eaten as roasted, fried, boiled, boiled and pounded; it can be cut and dried into chips and powdered, etc.

*Corresponding author: Dr. Rabirou Kassali, Department of Agricultural Economics, Obafemi Awolowo University, Ile-Ife, Osun State, Nigeria. Zip code: 220282, Ife Central. E-mail: rkassali@oauife.edu.ng; kasskassali@yahoo.com

There are several cultivars of yams according to different areas of the world, *D. rotundata Poir* and *D. Canyenensis Lam* are the most cultivated varieties in Nigeria, with *D. rotundata Poir* being the most preferred variety (Opara, 2003). In Nigeria, yam is cultivated in the guinea savannah and forest zone of the country besides cassava and maize. Major yams producing states include; Benue, Nasarawa, Niger, Oyo, Abia, Ekiti, Ondo, Taraba, FCT, Kaduna, Adamawa and Cross River States. Yam consumption spread across all the sphere of the country and therefore plays an important role in national food security, besides its cultural and spiritual role for the producing communities. Though produced essentially in the middle belt and South-Western zones of the country, it is traded to other parts of the country, especially, the far Northern States, and even exported to neighboring countries, like Niger Republic. The Federal Government of Nigeria, through the Federal Ministry of Agriculture and Rural Development is promoting its exports to more countries of the world, especially Europe and USA.

Yam is a bulky semi-perishable commodity, whose trade requires a strong transportation system, for it to be moved to distant urban centers and zones. In the rural markets of producing areas, yam occupies the position of "golden crop", as its seasonal production rhythms the market life, with all marketing activities attached to yam season. The presence of yam in the market determines the level of other activities in the market. Its marketing being the major source of cash and employment for majority of rural dwellers including farmers and other artisans and traders in the areas. Yam harvest season starts in most areas in August to reach a peak in November towards December, January and February.

In Nigeria, there is growing market gap between demand and supply as a result of fast growing population and unstable yearly supply. The situation is also aggravated by the lack of efficient marketing system, poor marketing performance and other inadequate storage facilities. Storage losses and marketing are major market challenges of yam business, besides corruption, bad roads, poor transportation system, and high cost of transport, price instability and poor of storage infrastructure in market places. In view of the current challenges it is therefore imperative to know how efficiently yam marketing is being performed in the study areas. The goal of an efficient marketing system being to ensuring low distribution cost, while guaranteeing better price to producer, which is also a way to sustain production.

The continuous increase in the demand for yam could be attributed to the value of the product and ever-increasing population resorting in product shortage and soaring market price. Other reasons could be due to production constraints such as root rot disease, scarcity of labour, inefficient marketing system and lack of capital (Izekor and Olumese, 2010). It is against this backdrop that this study analyzes the economics of yam marketing in the study area.

Statement of the research problem

Yam production and marketing play a significant role in the economy of Nigeria. Farmers frequently consider marketing as being their major challenge. However, while they are able to identify such problems as poor prices, market price instability, lack of transport and high post-harvest losses, traders and others cannot make investments in a climate of arbitrary government policy, poor roads, and increase in the cost of doing business, reduced return to farmers and increased prices to consumers. Adding to these, the growing problem of corruption that impacts negatively on agricultural marketing efficiency; increasing transaction costs for those actors in the marketing value chain. As a bulky agricultural commodity, wholesalers and retailers play an important role in yam marketing. These institutions perform functions including; assembling, storage, transportation, standardization, financing and risk bearing. The broad objective of the study is to assess the marketing of yam in Akoko North East Local Government Areas of Ondo State. Specifically, it: (i) determines the socio – economic characteristics of yam marketers; (ii) Analyzes the structure, conduct and performance of yam marketing; (iii) identifies the problems facing yam marketing in the study area.

METHODOLOGY

Study Area

The study was carried out in Akoko North-East Local Government Area of Ondo State, Nigeria. Its headquarters is in the town of Ikare, and has an area of 372km²with a population of 175,409 (NPC, 2006). The estimated population of the study area using the annual projected growth rate of 2.8% for the year 2017 is 234,698 inhabitants. Ikare is located in Ondo North Senatorial district at about 100km from Akure, the State Capital. The climate of the area is Savannah type, with two seasons, with the wet season spanning from March to October while the dry season covers late October to March. Each of these seasons is characterized by the influence of the South Westerly wind from the equatorial rain belt (NIMET, 2013).

In Ondo State, yam is grown extensively for various uses. In Ikare, yam has become the main staple food where the tuber is either boiled and eaten directly or pounded to form a cherished delicacy called pounded yam ("Iyan"). To stress the cultural importance of the commodity, a popular "New-Yam" Festival in Ikare is celebrated every June 20th which is done for no other crop. The market of yam is also being handled by wholesalers and retailers.

Sample Size and Sampling Technique

For this study, Two-stage Random Sampling Technique was adopted to select the respondents. The first stage was the purposive selection of three main yam markets in the area. The second stage involved the selection of 10 yam wholesalers and 20 yam retailers respectively from each market, making a total of 30 wholesalers and 60 retailers for the study.

Data Collection

Primary and secondary sources of data were used to achieve the objectives of this study. The primary data collected include; quantity of yam bought, selling price of yam, loading and offloading charges, transportation cost, rent, market tax, depreciation and product losses.

Data analysis

Descriptive statistics such as arithmetic mean, standard deviation, frequency distribution and percentage were used for analysis.

Market Structure Analysis

This was done using the following approaches:
- **Concentration Ratio:** This parameter measures market structure using the ratio of the two, four and eight largest firm's sales to the total sales of all sampled as follows;

$CR_n = \sum_i^n s_i / S$
With n = 2; 4; and 8
s_i = i^{th} largest firm's sales
S = Total sales of all firms

- **Gini Coefficient:** The Gini Coefficient (GC) was used to determine the degree of competition or monopoly in the market. The model is specified as follow;
$GC = 1 - \sum XY$

Where,

GC = Gini Coefficient
$\sum$ = Summation
X = percentage distribution of sales
Y = cumulative percentage distribution of sales revenue

Market Conduct Analysis

This is one of the most important components of a comprehensive market behavior analysis. Market conduct analysis pave ways to assess competitors' strengths and weaknesses in market place and implement effective strategies to improve competitive advantage, assess the strengths and weaknesses and uncover the objectives and strategies in a given market segment.

Market Performance Analysis

a) **Marketing Margin**
MM = SP - BP

Where, MM = Marketing Margin
SP = selling price
BP = buying price

b) **Marketing Profit (100 tubers)**
Profit = MM - MC
Where,
MC = Marketing Cost = cost of transport, handling, marketing charges, tax, shop rent, loading and offloading costs.

Rate of return = $\frac{(MM-MC) x 100}{BP \mp MC}$

c) **Markup Analysis**
Mark up = $\frac{(SP-BP) \, x \, 100}{SP}$

d) **Operational Efficiency (OE) Analysis**
$OE_i = \frac{sales}{MC}$ (Local efficiency)

$OE = \frac{OE_i}{OE_0} \times 100$ (Global efficiency)

OE_0 = Most locally efficient firm

RESULTS AND DISCUSSION

Social-Economic characteristics of the respondents

The Results in Table 1 revealed that yam marketers in all three markets visited were female. This showed that yam marketing seems solely a female business in the area, which confirms Oladapo *et al.* (2015) and Okoedo-Okojie and Okwuokenye (2016), but shows difference with (Okwuokenye and Onemolease (2011) and Bekun (2017) that reported mixed distribution in equal proportions in their various studies at Delta State, Nigeria and Bosso Local Government Area of Niger State, Nigeria. The relative sole distribution may be as a result of social barriers. The age distribution of respondents as presented in Table 1 indicates that 40% of wholesalers are between ages 51 and 60, while 41.7% of retailers are within the range of 41 to 50, meaning that most yam marketers in the study area are in their middle age. The average age was estimated at 50 years for wholesalers and 40 years for retailers. This result runs opposite to Hamidu *et al.* (2014) and Bekun (2017) that reported middle age to younger traders, in their respective areas.

It was revealed that 53.3% of wholesalers are married as against 66.7% for retailers. 13.3% and 18.3% were observed as single for both wholesalers and retailers respectively, while 33.3% of wholesaler and 15.0% of retailers are widows. The result shows that most of the

Table 1: Socio – economic Characteristics of yam marketers

Variable	Frequency		Proportion	
	Wholesalers	Retailers	Wholesalers	Retailer
Gender				
Male	0	0	0	0
Female	30	60	100	100
Age (years)				
20 - 30	04	11	13.2	18.4
31 – 40	02	11	6.6	18.4
41 – 50	09	25	30.0	41.7
51 – 60	12	09	40.0	15.1
61 – 70	03	04	10.0	6.7
Marital Status				
Married	16	40	53.3	66.7
Single	04	11	13.3	18.3
Widowed	10	09	33.3	15.0
Education qualification				
Primary	05	08	16.7	13.3
Secondary	09	21	30.0	35.0
Tertiary	00	02	0.00	03.3
Others	16	29	53.3	48.3
Household size				
1 – 5	14	02	46.7	03.4
6 – 10	16	30	53.3	50.0
11 – 15	00	28	00	46.6
Years of experience				
1 – 10	03	17	10.0	28.4
11 - 20	08	22	26.7	36.8
21 – 30	12	14	40.0	23.4
31—50	07	07	23.3	11.7
Use of labour				
Yes	24	35	80.0	58.3
No	06	25	20.0	41.7
Number of Shops				
1—2	26	59	86.7	98.3
3—4	04	01	13.3	1.7
Tuber lost per week				
0—5	16	53	53.3	88.4
6—10	13	07	43.4	11.7
11—20	01	00	03	00

Source: Data analysis, 2015

yam marketers in the study area are married and therefore yam marketing would serve as a means to meet the needs of the family. Table 1 also revealed that 53.3% of the wholesalers had non-formal education; 16.7% had primary education, while 30.0% had secondary education. For the retailers, 48.3% had non-formal education, while 13.3%, 35.6% and 3.3% had Primary education, Secondary school education and Tertiary education respectively. This result reveals that majority of yam marketers in the study area do not have formal education especially, wholesalers with high level of illiteracy.

In terms of household size, the results also indicate that 46.7% of wholesalers had household size of between 1 and 5 inhabitants, with 53.3% belonging to the household size of between 6 and 10 persons, while this is 3.4%, 50.0% and 46.6% respectively for retailers. The relative

size of the household implies availability of labour which could be an opportunity for the yam business. The study further revealed that 10% of the wholesalers have been in the business for about 10years, whilst26.6% (11-20years), 40% (21-30 years) and the remaining 23.3% (31-50 years). On the other hand, 28.4% of the retailers have been in yam business for same 10years as the wholesalers. The wholesalers that are in the category of markets with experience of between 21 and 30 years accounted for 36.8%, while those marketers with experience of between21 and 30years and 31 to 50 years attracted 23.4% and 11.7% respectively. The average experience of wholesalers was determined at 21years while that of retailers was estimated at 11 years. This implies that the wholesalers had more experience than the retailers in the yam business. Likewise, results in Table1 indicates that 80% of the wholesalers employed hired labour, while 20% did not, whilst58.3% of the retailers used labour to market yam against 41% who did not. This further demonstrates that the wholesalers used more of hired labour than the retailers. For shop ownership, 86.7% of the wholesalers owned a maximum of 2 shops to market their business, while 13.3% operates their yam marketing in 3 to 4 shops. On the other hand, 98.3% of the retailers employs a maximum of 2 shops as against a marginal value of 1.7% had 3 - 4 shops to trade yam. The reason for more than one shop is to improve sales and market share using location advantage. In terms of losses, the study revealed that 53.3 % of wholesalers lost about 5 tubers; 43.4% between 6 and 10 and 3.3% between 11 – 20 tubers. While on the side of the retailers, 88.4% of them lost about 5 tubers weekly, with those within the 6 to 10 tubers lost accounted for 11.7%. Most losses were as a result of improper storage.

Market Structure Analysis

Market Concentration

From the result presented in Table 2, the 2-firms concentration ratio for wholesalers showed 0.15%, while that of retailers was calculated at 0.089%. In the case of the 4-firms concentration ratio, the wholesalers had 0.257% while the retails showed 0.154%. The 8-firms ratio of 0.338% and 0.264% were estimated for wholesalers and retailers respectively. These results are indications of a competitive yam market both at wholesale and retail levels.

Table 2: Measure of concentration ratios of yam marketing in the study area

Concentration ratio	CR2	CR4	CR8
Wholesalers	0.15	0.257	0.338
Retailers	0.089	0.154	0.264

Source: Data analysis, 2015

Gini Coefficient Analysis

Table 3(a) and 3(b) show the Gini coefficient values for wholesalers and retailers of 0.3163 and 0.307 respectively. The results indicate high level of competition among yam marketers. These results agree with Ada-Okungbowa (2006) and Anuebunwa (2002) on yam marketing in Ondo and Abia States respectively; meaning there is high degree of competition in yam marketing in the study area. But this contradicts the findings of Reuben and Mshelia (2011) in Taraba State though and possibly due to location differences as they reported non-competitive wholesale and retail yam markets.

Table 3(a). Gini Coefficient analysis for wholesalers

Range (Millions)	Frequency	Percentage	Cumulative Percentage	XY
1-2	05	0.166	0.166	0.0275
2.1- 4	16	0.533	0.699	0.3725
4.1- 6	07	0.233	0.932	0.2171
6.1- 8	02	0.666	0.100	0.0666

Source: Field data analysis, 2015. M= million; $\sum XY$ (WS) =0.6837; G.C (Wholesaler) = 1 - 0.6837; G.C = 0.316

Table 3(b): Gini Coefficient analysis for retailers

Range (Millions)	Frequency	Percentage	Cumulative percentage	XY
0.1-0.5	09	0.15	0.15	0.0225
0.6 -1	31	0.516	0.666	0.3436
1.1- 2	19	0.316	0.982	0.3103
2.1- 4	01	0.066	0.100	0.0166

Source: Field data analysis, 2015   M =million; $\sum XY$ (RT) =0.6930; GC (Retailer) = 1 – 0.693; G.C = 0.307

Market Conduct in the Study Area

The survey of yam market in the area reveals total absence of predatory pricing or price collusion as price is mainly determined by cost and relatively by the market forces of supply and demand. Apart from sorting out yam by size and use of open display in the market, no advertisement or packaging are used. The result conforms to Folayan (2013)'s findings in Ekiti State.

Marketing Channel of Yam in the Study Area

The marketing channel reveals the flow of yam in the study area. The most frequently used channels for yam marketing as revealed by the study is reflected in Figure1.Channel 1 was from farmer (producers) to consumers through local assemblers and wholesalers to retailers, who finally sell to consumers. The other channel type observed in the area was from producer directly to speculators who act as agents between the farmer and the consumer. The consumers always detest this channel due to the sharp practices indulged in by inflating the price of yam, making it difficult to get yam at affordable rate. The analysis compared favorably with the decentralized type of channel whereby yam moves from the producers through rural assemblers, then through wholesalers to the final consumers (Ada-Okungbowa, 2006).

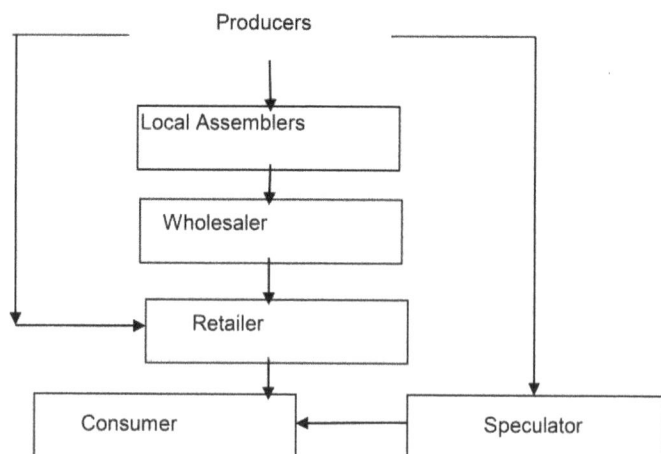

Figure 1: Yam marketing channels in the study area
Source: Field survey, 2015

Marketing Performance

Marketing profit (100tubers)

The marketing profit as determined from the analysis is based on one hundred (100 tubers) and this is reflected in Table 4.

Table 4: Market profit analysis

Item	Mean Value	
	Wholesaler	Retailer
(a) Selling price (SP)	17750	16433
(b) Buying price (BP)	14666	14008
(c) Marketing Margin(MM)	3084	2425
(d) Marketing cost (MC)	1074	787
(e) Profit= MM—MC	2010	1638
Rate of return	12.8%	11.0%

Source: Field data analysis, 2015

From the analysis, it is observed that the profit firm the yam business in the study area was estimated by subtracting the marketing cost (MC) from the Marketing margin (MM). For convenience and to reflect the market practice in the area, 100 tubers of yam was considered in the calculations. The result shows that the profit realized by the wholesalers was N2,010, while that of retailers was captured at N1,638. This further reveals the percentage of rate of returns per Naira invested for the wholesalers and retailers were 12.8% and 11.0% respectively.

Pricing efficiency of yam market in the study area

Table 5: Pricing efficiency analysis

	Wholesaler			Retailer		
Variables	Mean	SE	Sig.	Mean	SE	Sig.
MM	3038.3	200.6	.555	2458.33	100.09	.326
MC	1074.4	49.407		787.553	48.497	
t-value	.59[NS]			-0.99[NS]		

Source: Data analysis, 2015

Table 5 shows the analyzed result of yam market pricing efficiency in the study area The non-significant t-test value of 0.598 for wholesalers and -0.990 for retailers indicates that there is no difference between marketing margin and marketing cost at both wholesale and retail levels, meaning the existence of pricing efficiency in yam marketing in the area, meaning prices at both wholesale and retail markets reflect cost of marketing and that yam marketers behaviors in the area are not exploitative.

Operational efficiency of yam market in the area

The result presented in Table 6captured the operational efficiency of wholesaler. It shows that 19.8% Of the yam markets are within operational efficiency of 70 to 80, with 62.7% operated between 81 to 90, while those in the efficiency range of between 91and 100 accounted for 16.5%. It is equally noted from the analysis that the operational efficiency of the retailers within same categories of the wholesalers accounted for 1.7%, 74.8% and 25.5% respectively. Majority of both wholesalers and retailers falls within the efficiency range of 80-90 %. In comparing the efficiencies of the wholesalers and retailers, the retailers' shows higher operational efficiency than wholesaler, except at range 70-80 where wholesalers had 19.8% as against 1.7% for retailers. Overall, both wholesaling and retailing functions are operationally efficient. This means that these functions of marketing were performed at the lowest costs possible in the area.

Table 6: Operational efficiency analysis

Range	Frequency		Percentage	
(percentage)	Wholesaler	Retailer	Wholesaler	Retailer
70 - 80	06	01	19.8	1.7
81 - 90	19	44	62.7	74.8
91 - 100	05	15	16.5	25.5

Source: Data analysis, 2015

Markup analysis

Markup is the percentage added to the cost of goods to obtain selling price.

$$Markup = \frac{SP - BP - MC \times 100}{SP}$$

$$\text{Markup value for Wholesalers} = \frac{17750 - 14666 - 1074 \times 100}{17750}$$

$$= 11.3\%$$

$$\text{Markup value for Retailers} = \frac{16433 - 14008 - 787 \times 100}{16433} = 9.9\%$$

From the above, it is seen that the markup price for wholesalers and retailers were estimated at 11.3% and 9.9% respectively, which is a sign of reasonable pricing of yam in the area, in view of supply coming from far places.

Constraints faced by yam marketers in the area

Problems associated with yam marketing in the study is analyzed and captured in Table 7. From the table, it is seen that the wholesalers are faced with insecurity challenges (33.3%) as the most serious constraint, followed by price uncertainty (16.7%) and high cost of yam (13.3%). The Retailers reported high cost of yam (26.7%) as the most serious constraint followed by price uncertainty (23.3%), capital intensive (16.7%) and fragile nature of yam (13.3%) as the major constraints, besides high cost of rent (8.3%), insecurity (8.3%) and high cost of transportation (3.4%) respectively. Similar findings were also reported by Folayan (2013) in Ekiti State.

Table 7: Constraints of yam marketing faced by wholesalers in the study area

Constraints	Frequency		Percentage	
	Wholesaler	Retailer	Wholesale	Retailer
High cost of rent	02	05	06	08.3
Capital intensive	03	10	10.0	16.7
High cost of yam	04	16	13.3	26.7
High transportation	03	02	10.0	03.4
Insecurity	10	05	33.3	08.3
Fragile nature of yam	03	08	10.0	13.3
Price uncertainty	05	14	16.7	23.3
Total	30	60	100.0	100.0

Source: Data analysis, 2015

CONCLUSION AND RECOMMENDATIONS

The comparative analysis of wholesalers and retailers showed difference in marketing cost, marketing margin, marketing profit and markup. The concentration ratio, for wholesalers and retailers revealed that the market is perfectly competitive and the business as reported is profitable for both wholesalers and retailers. The market is price and operationally efficient, while most constraints affecting the smooth running of the business include high cost of yam and price uncertainty at retail level and insecurity at wholesale level. Based on the findings, the following recommendations are drawn:

i. Government and stakeholders should help in the construction and development of market infrastructure for efficient and effective marketing system. This will serve as revenue booster to all tiers of government

ii. Good storage facility should be provided relevant stakeholders to reduce yam spoilage.

iii. Traders should be encouraged to form formidable co-operative groups for the purpose of accessing relevant inputs for the continuity and sustainability of their business.

iv. Rural roads network be provided and rehabilitated in the area in order to alleviate the problems of transportation; this will also help in stabilizing the price of the commodity.

REFERENCES

Ada-Okungbowa CL (2006). The Market Structure, Conduct and Performance for Yam in Ondo State, Nigeria. Agrosearch, 4 (1&2): 12-20.

Adegeye AJ, Dittoh JS (1985). Essentials of Agricultural Economics. Ibadan: Impact Publishers.251 pp.

Adekanye TO (1977). Market Structure for Foodstuff: Problems and Prospect for Rural Development in Nigeria. In: TO Adekanye (Ed.): Reading in Agricultural Marketing. Ibadan. Longman Nigeria, pp. 101 – 107.

Anuebunwa FO (2002). A Structural Analysis of Yam Trade Flows into Abia State of Nigeria. Nigerian Agricultural Journal, 33:17–22.

Atkinson FS, Foster-Powell K, Brand-Miller JC (2008). International Tables of GlycemicIndex and Glycemic Load Values: Diabetes Care. 31(12): 2281-2283. Retrieved from: https://doi.org/10.2337/dc08-1239

Babaleye T (2003). "West Africa improving yam production technology". ANB-BIA supplement issue/Edition Nr463.

Bekun FV (2017). Economics of Yam (Dioscoreaceae Dioscorea) Marketing: New Insights from Bosso Local Government Area of Niger State, Nigeria. Preprints 2017100092. (doi: 10.20944/preprints 201710.0092.v1). Retrieved from: https://www.preprints.org/manuscript/201710.0092/v1/download

Fasasi AR, Fasina OO (2005). Resource Use Efficiency in Yam Production. Proceeding of the 39th Conference of the Agricultural Society of Nigeria, October 9 – 23, 2005; 184 – 187.

Folayan, JA (2013). Assessment of structure and conduct of yam wholesale market in Efon Alaaye Local Government Area of Ekiti State, Nigeria. African Journal of Food Science. 7(2):14-24. Retrieved from: http://www.academicjournals.org/AJFS

Food and Agriculture Organization [FAO] (2011). Available online at: www.foastat.fao.org.

Food and Agriculture Organization [FAO] (2002). Food and Agricultural Organization Year Book. P. 56.

Huxley A ed. (1992). New RHS Dictionary of Gardening. Macmillan.

Hamidu K, Adamu Y, Mohammed SY (2014). Spatial Profit Differential of Yam Marketing in Gombe Metropolis, Gombe State, Nigeria. Journal of Biology, Agriculture and Health Care. 4(12):72-76.

Izekor OB, Olumese MI (2010). Determinants of yam production and profitability in Edo State, Nigeria. African Journal of General Agriculture. 6(4): 205 – 210. Retrieved from:http://www.asopah.org

John T (2003). White [1] planning and designing rural markets, FAO Rome.

Kay DE (1987). Root Crops. Tropical Development and Research Institute: London

Marion BW, HardyCR (2004). Marketing Performance: Concept and Measure, Agricultural Economic Report. No. 244, Economic Research Service USDA Washington, D.C. pp. 11.

Mbah SO (2010). Analysis of Factors Affecting Yam Production in Ngor-Okpala Local Government Area of Imo State. In J. A. Akinlade; A. B. Ogunwale; V. O. Asaolu; O. A. Aderinola; O. O. Ojebiyi; T. A. Rafus; T. B. Olayeni, O. T. Yekinni
(Eds). Proceedings of the 44th Annual Conference of the Agricultural Society of Nigeria, LAUTECH, Ogbomoso, 18th - 22th October, 2010. Pp 340-344.

National Population Commission [NPC] (2006). National Population Census figures, NPC Bulletin.

NIMET (2013). Seasonal Rainfall Prediction. Retrieved from: *nimet.gov.ng/seasonal-rainfall prediction.* Pp. 1–20.

Okoedo-Okojie DU, Okwuokenye GF (2016). Characteristics and Potentials of Retail Marketing of Yam in Delta State, Nigeria: Implications for the Extension Services. British Journal of Applied Science & Technology. 14(2):1-8.

Okwuokenye GF, Onemolease EA (2011). Influence of socio-economic characteristics of yam sellers on marketing margins among yam wholesalers in Delta State, Nigeria. Journal of Agriculture and Social Research (JASR).
11 (1):81-90.

Oladapo MO, Osundare OT, Osundare FO, Oyebamiji K (2015). Effects of infrastructure on yam tuber Marketing in four selected local government areas of Ekiti State, Nigeria. Academia Journal of Biotechnology 3(6): 097-103.

Olukosi JO, Isitor SU (2002). Introduction to Agricultural Marketing and Prices: Principles and Application. Living Books series, G.U Publication, Abuja F.C.T, Nigeria.

Opara LU (2003). Yams: Post-Harvest Operations. FAO/Massey University, Private Bag 11-222, Palmerston North, New Zealand. 22 pp.

Oreri ST (2004). The Structure, Conduct and Performance of the Yam Marketing System in Benue State. Unpublished M.Sc. Thesis, Department of Agricultural Economics and Rural Sociology, A.B.U Zaria.

Pius CL, Odjurwuedemie EL (2006). Determinants of Yam Production and Economic Efficiency among Small Holders Farmers in Southern Nigeria. J. Central Eur. Agric. 7(2):33-342.

Reuben J, Mshelia SI (2011). Structural Analysis of Yam Markets in Southern Part of Taraba State, Nigeria. J. Agric. Sci, 2(1): 39-44.

7

Sharing gains of the potato in Kenya: A case of thin governance

Joseph Gichuru Wang'ombe[1*] and Meine Pieter van Dijk[2, 3]

[1] African Population and Health Research Center, P.O. Box 10787, Nairobi 00100, Kenya
[2] Maastricht School of Management, Maastricht, Netherlands
[3] UNESCO-IHE Institute for Water education, Delft, Netherlands

The potato offers a good alternative for diversification from maize, the staple food in Kenya. This article presents the results of a study on the potato marketing system, the factors affecting prices and the predominant governance system impacting on the market. Survey data were collected from 402 farmers in the three potato growing regions and addition information on monthly prices in major markets was provided the Ministry of Agriculture. There were also semi-structured interviews with the major actors in the potato sector.
Potato marketing exhibits the captive governance structure with traders collaboratively acting as the lead firm. An analysis of the split in selling price between the various players indicates that margins were concentrated at the coordination of marketing activities as opposed to the production activities. Coordination activities took about 40% of the consumer price. Besides the market channel used, the production region, the size of the land cultivated and the yields obtained determined the price obtained by the farmers. Given the distribution of the value added in the chain and the current dominance of traders, we argue in favour of upgrading the value chain and giving more power to the farmers and their organizations.

Key words: Marketing, value chain, prices, governance

INTRODUCTION

The leading food crop in Kenya in terms of annual tonnage is maize followed by potatoes and beans (Republic of Kenya, 2011). Maize is the main staple food in Kenya but its role in feeding the population is now greatly challenged as famine is constantly experienced in many regions of the country. In Kenya, the supply of maize and its price is usually a political issue as insufficient maize supply is synonymous with famine as country relies on imports with attendant higher costs. The potato offers a good alternative for diversifying from maize where Kenya is a net importer. It ranks next to maize in tonnage but is yet to assume an important place in food tables especially in rural areas where the crop is not grown. The potato is, however, in great demand by urban residents where it is made into potato chips (French fries) in hotels and served as part of stew at the household level.

Potatoes are grown in the highlands of Kenya. The main potato growing areas are between 1,400 and 2,700 metres above sea level with mean annual rainfalls of 1,000 mm or greater (Durr and Lorenzl, 1980).

*Corresponding author: Dr. Joseph Gichuru Wang'ombe, African Population and Health Research Center, P.O. Box 10787, Nairobi 00100, Kenya. Email: jgichuru@aphrc.org

Areas receiving more than 1,000 mm of rainfall per year are classified as high rainfall zone and occupy less than 20% of the productive agricultural land in Kenya, but carry about 50% of the country's population (Republic of Kenya, 2010). Potatoes therefore grow in the high altitude, high rainfall zone and the most suitable areas for production are:

1. Meru, Embu, and Kirinyaga Counties on the slope of Mt. Kenya;
2. Nyeri, Murang'a, Kiambu, and Nyandarua Counties on both sides of the Aberdare Range;
3. Nakuru and Kericho Counties along the Mau range; and
4. Several highland areas in Nyanza and Western Regions including Nandi, Uasin Gishu, Kakamega, Kisii, Bungoma, Busia, and Trans-Nzoia.

Though Nyanza and Western regions are suitable for potato production, data aggregated by regions indicate that the two produce insignificant quantities as the Central region leads in production followed by the Eastern and the Rift Valley regions (Republic of Kenya, 2007).

Potato consumption in urban areas is increasing. Urbanization has stimulated a dietary transition from traditional staple foods towards convenience foods, animal protein, fresh fruits and vegetables (Horton, 2008). Packaged foods are becoming more popular with the increasing trend of shopping in supermarkets and even fresh commodities like potatoes are being packed and marketed in supermarkets. Processed potatoes are greatly in demand by urban residents, in various forms as crisps and French fries. The urban market provides many value-adding opportunities including production of potato flour, which if exploited can help to enhance value addition.

Due to their bulkiness and perish ability, potatoes are largely consumed within the country and very little is exported. The best exporter in Africa exports less than 20% of its production (Food and Agriculture Organization (FAO), 2010). As such, prices in different countries are at different levels, since they are not influenced by external market forces. In Kenya, consumption potato prices are high and many people are not able to afford them. Potatoes in Kenya are considered a high quality and prestigious food, so that higher consumption rates are associated with higher incomes. Official statements in Kenya indicate that the potato is not considered a "poor person's food" (Ministry of Agriculture (MoA), 2007). The high prices of potatoes in Kenya is an indicator of the low level of development of the sector since more developed economies that produce potatoes present them as cheap food.

Despite relatively high prices that consumer have to pay, farmers have continued to complain about the poor returns from potatoes. Farmers' representatives lobbied the Ministry of Agriculture and the Ministry of Local Government to enact some laws to enhance fairness in the market place but the statutes though enacted have largely not been enforced. The Ministry of Agriculture, through Legal notice No. 44 of 2005 (GOK, 2005), legislated seed and ware potato production and marketing standards. The most celebrated part of the legislation was on the size of bag used to market potatoes. Seed potatoes were to be marketed in bags of 50 kilograms while ware potatoes were to be marketed in sisal or jute bag weighing 110 kilograms. To enable enforcement of recommended size of bag by the local councils that control markets, the Ministry of Local Government published Legal Notice no. 113 in 2008 (GOK, 2008) which gave local authorities powers to enforce the recommended packaging. These adoptive By-Laws oblige market players to use the recommended packing of various crops and impose penalties for failure to adhere to recommended weight and packaging. However, not much of the legislation has been enforced and marketing of potatoes has continued to be a challenge to farmers. Indeed, out of 396 farmers who ranked five constraints in the order of severity, 30% ranked marketing as the most serious constraint followed by seeds (22%), roads (21%), diseases (16%) and fertilisers (11%) (Wang'ombe, 2013). We therefore sought to find out; (1) the factors affecting the price obtained by the farmers at the farm gate, (2) how much of the final market price is retained by farmers, and (3) the governance arrangements that were in place and how these were affecting the value distribution.

Theoretical Framework

Kaplinsky (2004) summarizes the determinants of gains in the global market place. He first points out barriers to entry as determinants of distribution of rents. Those who are stuck in activities that have low barriers to entry in a world of increasing competition find themselves on a losing streak. Secondly, rents such as copyrights and brand names are increasingly found in intangible parts of the value chain. Thirdly, rents are larger in the coordination and control of value chain activities or governance. It therefore becomes important who controls the chain. Control is usually through the setting of parameters or requirement for participation and enforcement of agreed standards. In producer driven chains, the key parameters are set by firms that control key product and process technologies meaning that most of such chains are in capital and technology intensive industries. In buyer driven chains, the key parameters are set by retailers and brand name firms, which focus on design and marketing and usually do not possess any production technologies (Humphrey and Schmitz, 2002). According to Kaplinsky (2004), buyer driven chains have critical governing role played by the buyer while producer driven chains have the producers at the driving seat controlling vital technologies and playing the role of coordinating the various links. Buyer driven value chains as opposed to producer driven chains are characterized by complex forms of coordination and integration and rules of participation (Vorley, 2001).

Value chain governance activities are often subject to high entry barriers and hence provide higher returns. Governance activities usually extend outside the firm and demand resources for managing the various processes. To meet the standards required by the market, different relationships may develop between players in the market. Kaplinsky (2004) looked at governance in three ways; legislative (setting the rules for participation), judicial (auditing performance, checking compliance with the rules) and executive governance (proactive assistance in order to meet the operating rules).Standards are set by internal or external parties. We therefore have self-regulation or external regulations where agents like the government set product and process parameters.

Gereffi et al. (2005) postulated three determinants of the type of governance as the relative complexity of transactions, the ability to codify or systemise transactions and the capability of the supply base. Based on these three determinants, they outlined five governance structures. These are (1) hierarchy, (2) captive, (3) relational, (4) modular and (5) market. The hierarchy governance structures are characterised by complexity of transactions, thereby requiring vertical linkages and managerial control of supply processes from the head office of an integrated firm, as subordinate office is seen not to have the capability to systemise processes without management control. The captive structure is similar with the difference being that monitoring and control is by a lead firm who invest in enhancing the capacity of the small firm to supply solely to themselves. The switching costs are high as the lead firm support will be lost on switching and the processes are too complex for the small firm to handle alone. The suppliers therefore remain captive. Relational governance structure demands close relationship between the lead firm and the suppliers which creates mutual dependence and a high level of specificity in the product, since there is mutual understanding and requisite capacity. In modular governance structures, both suppliers and buyers work with multiple partners. The suppliers have good capacity to make product to customer specifications. Finally, the market governance structure has transactions or product specifications that are easily codified by the many suppliers. Buyers can buy from any player in the market and supplier equally can serve any customer.

Humphrey and Schmitz (2001) came up with 4 types of governance which are very similar to Gereffi et al. (2005). Their first type was 'Arm's Length' which can be equated to Gereffi's market governance structure. In this mode of governance, buyers and suppliers are not in a close relationship. The buyers do not give a special commitment to their partners. His second categorisation was 'Networks'. Here global buyers cooperate in an interdependent relationship with their suppliers where they share competences. Depending on the interdependence, this governance type can either be equated to modular or relational in Gereffi's categorisation. The third one was 'Quasi-hierarchy'. Here, global buyers control the operations in the chain by specifying the characteristics of the desired products and sometimes the processes to be followed and controlled. Parties become subordinate such as through sub-contracting relationship. This resembles the captive structure in Gereffi's. The last was 'Hierarchy' which is exactly the same as Gereffi's categorisation under the same name. Here, global buyers control the operations of the chain by controlling the ownership.

The length of the chain can determine the amount that is retained by the various participants. Trienekens (2011) postulated that relatively long chains have value distributed over a large number of participants and also longer transportation both in terms of distance and time. Final market information rarely get transmitted throughout the chain in case of longer chains and most producers end up selling an undifferentiated product where little value addition has taken place e.g. local staple food chains in developing countries. As sophistication increases, the chain shortens with communication of final market requirements becoming critical. The three determinants of governance structure as per Gereffi et al., 2005 come into play: the relative complexity of transactions, the ability to codify or systemise transactions and the capability of the supply base.

The local stable food chains have a simple governance structure which Ruben et al. (2007) referred to as A-System which deals with an undifferentiated product supplied by numerous small scale producers to the local market. Ruben named the middle system B-System which has medium sized producers who are relatively organized, produce according to some minimal standards and supply to relatively organized markets like supermarkets. The final classification, C-System is now more complex and is typical of an export chain where producers and buyers are larger and more integrated and quality requirements are strictly adhered to.

The price offering in the market place is rarely stable for agriculture commodities. Price volatility results from a mismatch between supply and demand. The volatility results in periods of gains and losses especially for the farmer since their price offering is governed by the market conditions- high demand, low supply, high price. The converse is true. More predictability (low volatility) will reduce traders' risks allowing them to accept lower margins thereby providing consumers and farmers with better prices. Governance however changes the behaviour of the market. The demand and supply get redefined as rules of participation are set and only those meeting the rules of participation are included in the supply. In some cases, prices and quantities are predetermined. The study of prices and distribution of gains is therefore incomplete without incorporating governance issues.

Humphrey and Schmitz (2002) clearly demonstrated why governance matter. According Humphrey and Schmitz (2002), governance determines who is able to access the market and how the gains are distributed in line with power relationships. Governance also enables quicker acquisition of production capabilities especially in cases where the market requirements are stringent. Governance systems act as leverage points for policy initiatives e.g. safety standards. The governance system also renders itself for use in technical assistance.

METHODOLOGY

As part of a broader survey, farm gate prices and marketing channels data were collected from 402 farmers in the three main potato growing areas of Kenya corresponding to Nyandarua, Nakuru and Meru Counties. Data were also collected on household characteristics and adoption of various inputs and marketing innovations. The prices obtained at the farm gate were computed for a standard bag of 110 kilograms. Most farmers sold in extended bags weighing more than 110kgs but conversion was done to bring these to the price of standard weight bag. Monthly market prices at the main Nairobi market were obtained from the Ministry of Agriculture which collects daily market prices. The prices were then compared to the data obtained from the farmers through the survey while other information on transport were obtained from interviews with farmers and traders. The data presented on market prices only relates to the months in which the survey data were collected in order to be able to relate these to the farm gate price.

In each of the regions studied, group discussions with farmers were used as a way of introducing the study. Discussions brought out general constraints that were specifically studied and analysed using the survey questionnaire. We also implemented semi-structured interviews with organizations that support potato growing. Most of the interviews were with Ministry of Agriculture officers at the Ministry Headquarters and at the districts studied. The Ministry is the main agent providing executive governance (for the farmers to produce according to certain standards or rules) and is responsible for 79.55% of the visits made to the farmers in the last three years (Wang'ombe, 2013).

RESULTS AND DISCUSSIONS

Marketing of Potatoes in Kenya

Out of the 396 farmers who ranked five constraints in order of severity, 30% ranked marketing as the most serious constraint. This was followed by seeds (22%), roads (21%), diseases (16%) and fertilisers (11%). In the eyes of the farmer, marketing is a big issue. Roads which also rank high are indirectly a marketing problem as they influence the farm gate price since buyers have to factor in transport costs in establishing the price to offer. The farmers take into account the way the market operates as they evaluate possible innovations they can adopt. When faced with very low farm gate prices, farmers are unlikely to adopt expensive innovations since the market cannot provide enough income to pay back for the costs incurred.

Potatoes for consumption are stored for up to three months which is generally insufficient to maintain a supply of potatoes until the next harvest. Studies indicate that farmers bring potatoes straight to the market due to lack of storage facilities (Durr and Lorenzl, 1980; Kaguongo et al., 2008). The potatoes therefore end up flooding the market at harvest time resulting in depressed prices.

From the survey, we established that 90% of the farmers sell to middlemen or brokers. The marketing system is dominated by a large number of intermediaries who introduce the finance needed to get the crop out of the farm. The buyers in Kenya provide the capital for harvest and post-harvest activities. They pay the local middlemen to assemble a defined quantity of bags for transportation to Nairobi. The local middlemen will purchase packing sacks, transport produce from the farm and pay the farmer a spot price. In some cases, the spot price is adjusted to also cater for harvesting costs paid by the buyer. Rural brokers are directed by traders (who often own trucks which transport potatoes) to assemble potatoes at an agreed price several days prior to a trip. Brokers are responsible to traders for maintaining quality standards and correct quantities, i.e. checking that bags are properly filled. The buyer caters for all the levies on the road and take care of all the risks on the product. Because of the well-developed mobile phone payment system in Kenya, most of the monetary transactions happen by way of Mpesa transfers (mobile money transfers). The buyer in Nairobi sends by Mpesa the amount required for the product and pays all ancillary costs to be incurred up to the Nairobi market.

At central markets, such as in Nairobi, market brokers usually approach lorry operators/medium size brokers and offer to sell potatoes at negotiated prices and commissions. They sell to all manner of traders including wholesalers. Some of the wholesalers in the market are able to negotiate for full truck-load supply.

Smaller-scale agents distribute smaller quantities down the marketing line and provide services such as repackaging and hauling, ultimately on the backs of porters and hand pushed carts. Retailers buy from the market brokers who have a firm hold on the market. The trucks are not allowed to sell directly to the retailers. In Nairobi, fish and chips (French fries) restaurants buy several sacks a day. Depending on distance, these are

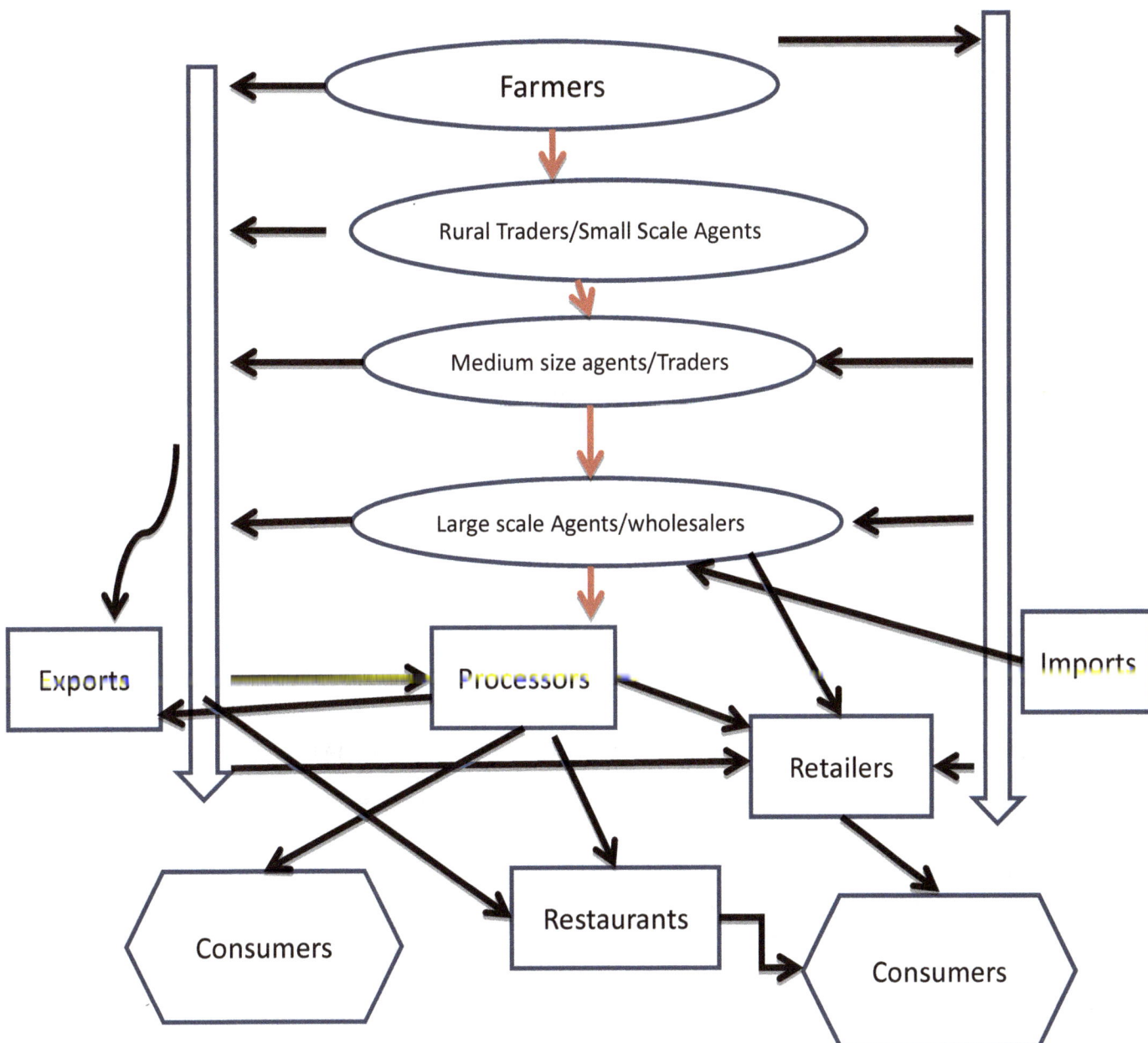

Figure 1. The Kenya Potato Value Chain

ferried by porters on their backs, in hand pushed carts or in pick-up vans. Larger hotels have entered into contracts with processors who buy potatoes from the market, process them into chips or French fries size and supply according to a defined schedule. The processors supply ready to fry potatoes in almost a just in time system as hotels prefer to take 1-2 days delivery requirements.

Some processors have direct contracts with farmers. Examples of such include Deepa Industries who process crisps for exports. The export contract demands a particular variety and control on quality which the open market cannot provide. Those selling crisps in the local market do not have very high standards to adhere to and

can afford to buy from the open market. The same applies to those processing for the middle sized restaurant. Those selling to five star hotels have to exercise more due diligence in their buying.

Consumers will mainly buy from wet markets in various regions using a 20 kilogram container ('debe') as a measure or smaller containers ('gorogoro'). Those who are not able to purchase quantities of 5 kilograms or above will mainly buy from small market stalls which can sell any quantity from one piece of potato.

The supermarkets in the last few years also started stocking potatoes. The measure they use is a kilogram and buyers will pack for themselves any quantity they

Table 1. Farm Gate, Nairobi Market Price and Margins

Data Collection Month	District	Potato Variety	Aver. Nairobi Prices	Aver. Farm Gate Price	Farm Gate as %Nairobi	Transport	Transport%	Trader Margin	Kenya's Inflation
Mar-11	Buuri	Red	4600	1629.48	35%	650	14%	50%	9.2%
		White	5100		32%	650	13%	55%	
Sep-10	Kuresoi	Red	3345	879.1	26%	800	24%	50%	3.20%
		White	3590		24%	800	22%	53%	
Jun-10	Nyandarua Central	Red	2233	908.72	41%	650	29%	30%	
Jun-10	Nyandarua South	White	2233	748.61	34%	650	29%	37%	3.50%
Jun-10	Nyandarua West			749.48	34%	650	29%	37%	
Jan-10	Mau Narok	Red	1560	617.7	40%	700	45%	16%	4.70%
		White	1460		42%	700	48%	10%	
Sep-09	Meru Central	Red	2710	1645.43	61%	450	17%	23%	6.70%
		White	2410		68%	450	19%	13%	
	Average				**40%**		**26%**	**34%**	

Source: Ministry of Agriculture, Kenya National Bureau of Statistics and Own research data

want. The supermarkets, in their desire to minimise the number of vendors selling to them, may appoint several vendors to supply vegetables to them and also include potatoes in the package. As more and more people buy their potatoes from the supermarket, efficient distribution mechanisms will arise including due diligence systems on quality. Currently, customers are not sure what potatoes to expect in the supermarket. Many therefore prefer the wet market where the choice is wider.

Sharing the Potato Price

Table 1 shows the different prices obtained at the farm level computed for a bag of 110 kilograms. The trader gross margin is taken as the difference between the farm gate price and transport costs to Nairobi which has been used as the common market for potatoes from all regions. The margin includes other costs traders have to meet like road and market levies plus their overheads.

Though final market prices have varied a great deal over time, the farm gate price has remained low. Table 1 shows that farmers retain 24-68% (average of 40%) of the market price dependent on the market chain followed, state of roads, varieties marketed and supply situations. Except for Meru Central, farm gate price of potatoes is less than 50% of the ultimate retail price. The data for Meru Central were collected in late 2009 when there had been drought in the country and farm gate prices were

high. The year 2010 had good rainfall and supply of the product was good. If we eliminate, Meru Central, we see that the farmers retain 24 to 42% which is a better representation of what happens in years with normal rainfall patterns.

Many factors can affect prices at all stages, including not only the relative supply and demand of potatoes, but also the availability of maize (often a factor of the rains), and the effective distance from a rural site to wholesale and retail markets coupled with the state of roads. High costs from the farm are indicators of poor rural roads in the production area while transport cost to Nairobi vary by the distance from the production area to Nairobi.

Trader's gross margins are very high with most of them at over 30%. The country's inflation over the study period was in single digit. Even after factoring other business overheads, the traders are likely to have earned super-normal profits. Durr and Lorenzl (1980) found net trade margins of 5 to 30% after accounting for marketing costs. The brokers or middlemen who control 90% of the market have put entry barriers and have continued to earn economic rent. A youth farmers' group in Nyandarua presented an experience of how they tried to sell their potato produce directly at the major Wakulima Market in Nairobi. They hired a lorry and transported a full load to the market. On reaching the market, they were blocked by a group of brokers who threatened them with dire action if they interacted with the buyers. They were forced to sell to the same brokers who they were trying to avoid at a price that was the farm gate plus transport costs. Their action yielded no positive results and they resigned to selling to brokers.

In the example above, middlemen collectively act as the captive lead firm but introduce little benefit to the chain. Lead firm conventionally support the chain by helping producers navigate complex processes. The lead firm in most cases also assures the producer of a certain price or price level. In the case of potatoes, prices fluctuate greatly and the only players who appear shielded from these effects are the brokers or middlemen who actually take advantage of the fluctuations. Farmers rarely gain as their prices remain depressed even when final market is giving better prices.

The figures in Table 1 are just rough indicators as prices and costs vary greatly. Supply and demand forces determine the price offering at Nairobi wholesalers market (Wakulima). When the price is good in the market, the traders retain a higher margin since they normally do not extend the extra gains to the farmers. This was the case in September 2010 and March 2011, when market prices were relatively high but the traders ended up retaining most of the gains as the farmers prices were not appropriately adjusted to correspond with the increased market prices. The traders retained 50% plus of the retail market price as per the rough analysis in Table 1. In times of glut, traders retain very little but they average their gains over time. Transport costs are influenced by rains as most of the potato growing areas have roads which can only be accessed by tractors and donkeys when it rains.

There is concentration of buyer power in potato marketing typical of buyer driven chains. Unlike the case of global value chains where large retailers, marketers and branded manufacturers play a pivotal role in setting parameters for participation, the middlemen have a lose form of collaboration which has resulted in some standards like the extended bag used to pack potatoes, buyer participation in packing, assembling and transporting to Nairobi. Any direct involvement of the farmer in these activities is met with stiff resistance by the traders. The capital provided for these activities is an important entry barrier. Potatoes are transported for long distances to the capital city, Nairobi such that transport costs constitute about 26% of the final market price. Other than transport costs, there are other costs like road cess charged by the local governments in addition to bribes often paid by transporters for smooth facilitation on the roads. The lack of capital by the farmers and their failure to collectively act together allow the traders to depict a captive governance structure where farmers are only able to access the market through them. Unfortunately, the buyers (middlemen) are not involved at all in enhancing the capacity of the farmers to supply potatoes. They only wait for the harvest and unlike other

captive structures, switching costs would not be high for farmers as the buyers are not involved in the complex production processes. The major constraint for the farmer is organizational where collective action would have produced the needed economies and facilitated acquisition of requisite capital for investment in marketing activities.

Factors Affecting Potato Prices

We now look at the factors that influence price in Table 2 below. The first part of the table presents the descriptive statistics that are used in the regression model. The second part presents the regression output on factors affecting potato prices.

Statistically significant regression results were obtained for production region, yields obtained, portion planted with potatoes, and those who sold to the local market and to private companies as opposed to brokers which was the reference category. A high R squared of 51.29 % was obtained. Though most of the potato growing areas have very poor road networks, distance to the nearest all weather road was not significant. The farmers, through the use of donkey-drawn carts are able to negotiate the bad roads from the farms as they deliver produce to a location where trucks are able to collect the product. However, the effect of the roads is perhaps captured by the region of production as the regions retain significantly different proportions of the sales price. Regions with lower transport costs retain a better proportion of the sale price. It is evident from Table 1 that Rift valley regions growing potatoes retained the lowest amount. Most of these areas have very delinquent transport system. For example, the researcher field trips to Mau Narok farms were mainly by motor cycle. He would pack his car at a place named *Mwishowa Lami*, a Kiswahili adjective for end of tarmac road and then hop into a motor cycle as the roads were very bad. Tractors transport potatoes up to Mwishowa Lami where trucks are waiting to load. Since brokers will have to pay the resultant high transport costs, they reduce the farm gate price considering that these potatoes will still have to compete with others from places where transport costs are lower.

The channel used to sell potatoes was only statistically significant for those who sold to the local market and to private companies (sold to brokers was the reference category). The coefficient for those who sold in the local market is negative while the coefficient for those who sold to private company is positive. This means that those selling to the local market obtained lower prices than those selling to brokers while those who sold to private companies obtained better prices than those who sold to brokers. Private companies usually establish contracts with farmers. Despite Kenya's rich history of contract farming for cash crops like tea and coffee, there are few

Table 2. Descriptive Statistics and Regression of Price

Descriptive Statistics	Mean	Std Dev.	Min	Max
Price per 110kg bag	1378.795	525.9758	500	3000
Distance to all weather road	2.55602	7.359155	0	100
Region - Central	0.3731343	0.48424	0	1
Region - Eastern	0.2985075	0.458174	0	1
Region- Rift Valley	0.3283582	0.470201	0	1
Yield	12.16793	6.310752	0.81543	33.16082
Sold to Brokers	0.910448	0.285895	0	1
Sold on Contract	0.034826	0.183567	0	1
Sold to Neighbours	0.029851	0.170388	0	1
Sold at the Local market	0.012438	0.110967	0	1
Sold to Private Company	0.012438	0.110967	0	1
Quantity sold	35.23664	46.96843	0	540
Weight of bag used to sell (Kgs)	144.2651	28.40757	1	180
Size of land owned	4.457916	5.873537	0	60
Membership of farmer group	0.310945	0.463457	0	1
Portion grown with potatoes	1.022077	0.775027	0.05	6

Regression on Price per 110 Kilogram bag	Coefficient	Robust Std. Err.	P value
Distance to all weather road	-2.19581	1.782957	0.219
Region - Central	-809.803	51.61397	0***
Region- Rift Valley	-730.114	57.14819	0***
Yield	10.75769	3.888134	.006***
Sold on Contract	157.5877	129.4507	0.224
Sold to Neighbours	-153.463	128.3158	0.233
Sold at the Local market	-522.61	80.73311	0***
Sold to Private Company	184.0241	50.51275	0***
Qty packed (less or equal to 110 kgs or greater than 110kgs)	-51.1334	49.84785	0.306
Size of land owned	-1.78876	2.914607	0.54
Membership of farmers group	-18.3109	43.92722	0.677
Portion planted with potatoes	47.79331	27.02739	0.078*
Constant	1768.136	71.76932	0***
R squared	51.29%		

*P <0.1, **P <0.05, ***P <0.01 (two tailed test).

reported contractual arrangement for potatoes. Examples include Deepa Industries and Njoro canners (Strohm and Hoeffler, 2006). Private companies set requirements for participation. Processing industries require potatoes that are not dirty, diseased, injured or have no deep eyes, are round and of a specific size and mature so that the potatoes develop a nice colour when fried (Strohm and Hoeffler, 2006). This necessitates some form of executive governance which has mainly been carried out by the

Ministry of Agriculture and NGOs. The low volumes moving through this chain do not justify private investment in executive governance.

Though most farmers sold their potatoes to middlemen or brokers, the percentage of those selling to middlemen varies by region. The Eastern region which retained the highest portion of the selling price also had the lowest percentage of farmers selling to middlemen at 82% as compared to 94% for Central and 95% for Rift Valley. The

Table 3. Descriptive statistics on sales for the different regions

To whom sold	Broker	%	Contract	%	Neighbour	Per cent	local market	%	Total	%
Region										
Central	140	94%	6	4%	3	2%	0	0%	149	100%
Eastern	93	82%	12	11%	4	4%	5	4%	114	100%
Rift valley	119	95%	1	1%	5	4%	0	0%	125	100%
Total	**352**	**91%**	**19**	**5%**	**12**	**3%**	**5**	**1%**	**388**	**100%**

Eastern region also had the highest percentage selling through contract at 11% as compared to 4% for Central region and 1% for the Rift Valley region. This is another important contributor to the significant results obtained for region both from the regression and as is evident in the secondary data from the Ministry of Agriculture as per Table 1. Different things are happening in different regions and there is a lot of scope for the regions to learn from each other.

Discussions with brokers indicated that quality standards are different for different regions. Brokers therefore have to focus on a particular source which they understand well in order to make good money. Due to the different standards and product varieties, prices in the same market are determined by source. For example, on the day the researcher was at the market, the potatoes from Molo were selling at Ksh. 1,000 above those from Njabini. Though the Molo potatoes are a different variety and one which is preferred for home consumption, the producers in Molo exhibit better due diligence by sorting out the potatoes before parking. The buyer said 'for *Molo, you are sure what you are dealing with. From Njabini, you will get a mixed bag and may have to sort before selling to buyers requiring higher standards like processors'*.

Those obtaining higher yields are also able to secure better prices. The most likely reason for this is that those getting higher yields are better endowed in other aspects. For example, we established that yields are influenced by the adoption of innovation and communication variables (Wang'ombe and van Dijk, 2013). These dynamics may also be related to negotiation for better prices. We have cases like Kisima Farm (Wangombe, 2013) where producer has sophisticated production system that delivers quality output which attracts buyers to the farm and the producer insists on selling using weight as opposed to smallholder farmers who just pack in a bag that is never weighed. The bags that smallholder farmers use deviate greatly from the recommended weight per bag of 110 kilograms with some weighing as much as 180 kilograms. Producers can therefore exercise power if they are able to demonstrate quality as a differentiator.
We unfortunately did not include a question on variety grown in the survey questionnaire. Earlier studies

indicate that market prices for potatoes are also dependent on market perception of particular attributes (Kaguongo *et al.*, 2008). Attributes considered by the market include the colour of the skin and the size of the tubers. The preferences are driven by taste and usage. For example, the French fries processors prefer white skinned potatoes and those with fewer noodles. Some consumers in certain regions prefer the red skinned potatoes due to their taste. Variety grown thus determine price. Meru Central grows Kerr's Pink and Ngure (red) also sometimes referred to as Meru. The quality of red skinned potatoes is rated very high by consumers. According to one of the national farmers' association official involved in our study, only Dutch Robijn comes close to red skinned potatoes in terms of quality for other uses besides French fries. Interesting, these varieties grown in Meru do not do well elsewhere as they suffers greatly from late blight when grown in other areas. This market niche makes Meru prices stable and relatively higher than other region. Earlier studies have indicated that Meru does much better than all the other regions (Durr and Lorenzl, 1980; Crissman *et al.*, 1993).

Legislative and Judicial Governance –Potato Production and Marketing

Legislation can have either positive or negative impact on a sector. Most legislation is driven by the need to control behaviour that may be seen to be undesirable. The effect of legislation can however be positive or negative depending on how the population and relevant stakeholders take it. For example, legislation banning sale of traditional alcoholic drinks which are still in great demand may make the brewers do their preparation in hidden and very unhygienic conditions thereby leading to health issues for the consumers. The legislation meant to curb consumption ends up being ineffective, as consumption continues at almost the same rate. Legislation defines the rules of participation and in most cases act as barriers to entry for those who are unable to comply.

The Ministry of Agriculture, through Legal notice No. 44 of 2005 (GOK, 2005), legislated seed and ware potato production and marketing standards. Included in the legislation was the requirement that seed potatoes be produced from such stocks as may be certified from time to time and bearing the characteristics stated in the second schedule to the rules. Pests and diseases were to be controlled by use of chemicals recommended by the Pest Control and Produce Board (PCMB). Also included in the legislation were rules on harvesting, storage and grading of potatoes. But the most celebrated part of the legislation was on the size of bag used to market potatoes. Seed potatoes were to be marketed in bags of 50 kilograms while ware potatoes were to be marketed in sisal or jute bag weighing 110 kilograms.

Our study found that out of the 393 farmers who responded to the question 'state the recommended weight of bag for selling the potatoes', 82% indicated the correct weight of 110 kilograms per bag. This is the same percentage that the Kenya National Potato Farmers Association (KENAPOFA) obtained in 2010. Though the awareness is high, our study revealed that 74% use extended bags or bags weighing more than 110kgs. Only 26% used the recommended weight including a 5% using other measures weighing less than 110 kilograms. The usage of recommended bags was lowest at 15% for Rift Valley farmers. It was 23% for Central with Eastern recording the highest percentage at 41%, meaning that there are great regional variances in the use of the recommended size of bag. KENAPOFA (2010) in their study on policy implementation and its economic impact on potato marketing value chain in Kenya found that only 12% were using the recommended weight. Their sample was however much smaller consisting of about 70 traders and 70 farmers.

Currently, middlemen provide the bags used to pack potatoes. They also pay for the transportation costs from the farm to the assembly site before the bags are loaded to vehicles destined mainly for the urban markets. The middlemen provide a bag made of the correct material (sisal or jute) which is one of the legislated requirements being followed. However, to get the bag to carry more than the recommended weight, an extension is added to the bag. We therefore end up with what is usually referred to as an extended bag. These extended bags weigh more than the 110 kg recommended weight by as much as 70%.

The use of the extended bag has a direct impact on returns for the farmers. Though regression as per Table 2 shows that the size of bag is not statistically significant in determining the price per kilogram, the regression coefficient is negative meaning that the higher the weight of the bag, the lower the price per kilogram sold. Apparently, traders just price sack loads without regard to the weight. The extended bag carries more potatoes (more weight) but its price is usually not different from a bag that is not extended.

To enable enforcement of recommended size of bag by the local councils that control markets, the Ministry of Local Government published Legal Notice no. 113 in 2008 (GOK, 2008), which gave local authorities powers to enforce the recommended packaging. These adoptive By-Laws have the recommended packing of various crops and impose penalties for failure to adhere to recommended weight and packaging.

Despite the enactment of the two legislations, the use of extended bag for marketing of potatoes has continued as shown by the high usage of extended bag which stood at 74% for the study. The use of extended bags denies farmers better price as they sell almost two bags for the price of one. There are also health impacts as the bags weighing sometimes up to 180kgs are transported and loaded on the backs of young men who after several years of doing this, experience back problems. This is one area where the government can proactively assist the farmers to be able to sell in the legislated bag size. Judicial governance is clearly needed to audit performance and check compliance to the legislated rules. Given that the agencies that are supposed to provide this judicial governance have failed to play the role, other actors can also come and fill this evident gap in the potato industry.

Are potato farmers getting a fair share of the price?

An efficient marketing system provides a fair price for farmers and low prices for consumers. When the farm gate is expressed as a percentage of the market price, comments are made on how fair or unfair the sharing of the market price has been. However, there is no particular yardstick to measure what is fair. For one to establish fairness, all activities undertaken by the various players in the value chain needs to priced and the difference between cost and value received computed as the gain of the particular player. For farmers, costs of production are routinely worked out. Purchased inputs and labor are easy to compute but the cost of self-supplied services is often difficult to value especially for small holder farmers since they do not keep records. The value of own supplied services may perhaps need to be valued at the opportunity cost which is usually lower than the actual cost of contracting external labour. Even considering incomplete records on costs of production, we still encounter instances where full cost recovery is not achieved leave alone profit. Though there is no objective yardstick for what is fair for the farmer, cost recovery should be the minimum standard.

The 60% plus that goes to potato traders and other processes may be seen as high, but there is need to consider the heavy investments they make in facilitating the processes required to get the product to the market. The farmers are ill prepared to invest in these processes. The farmers' organization has not even succeeded in bringing the produce together so that they can negotiate

Table 4. Region and Extended Bag (>110kg)

	110kg or Less	Extended (>110kg)	Total
Central	35	115	150
%	23.33	76.67	100
Eastern	49	71	120
%	40.83	59.17	100
Rift Valley	20	112	132
%	15.15	84.85	100
Total	**104**	**298**	**402**
%	**25.87**	**74.13**	**100**

Pearson chi2(2) = 22.4209   Pr = 0.000

Table 5. Cost Structure of African Fresh Fruit and Vegetables Exports to Europe Compared with Local Potato Value Chain

Sharing of Gains in Agriculture Export Chains

Value Chain	Producer	Packaging	Other processes	Export Market	Total	% producer in local market*	% Packaging*
Canned Peach - South Africa	12	12	19	57	100	28	28
Mangetout from Zimbambwe(exported to UK)	12	5	25	58	100	29	12
Fresh Vegetables from Kenya (Exported to UK)	15	13	20	52	100	31	27

*The denominator is the total less the export market
Source: C. Dolan, J. Humphrey, and C. Harriss-Pascal, "Value Chains and Upgrading; D.E. Kaplan and R. Kaplinsky, "Trade and Industrial Policy on an Uneven Playing Field: The Case of the Deciduous Fruit Canning Industry in South Africa

with the buyers. Coordination of the farmers is very poor and they lose some of the margin to this governance action by buyers. The farmers sell the product as is from the farm. They do not even sort out the product into various grades. As it were, the product is sold undifferentiated and another premium is lost. The materials for packing are provided by the buyer and so are the standards. The buyers insist on an extended bag which is used by over 70% of the farmers. If buyers would find potatoes packed in appropriate bags, it would be easier for them to accept the standard bags which weigh 110kgs as opposed to the extended bag of 130-180 kgs.

We can compare what potato producers in Kenya retain with some African agriculture food chains as below.

To be able to compare the local chain and the global chain in Table 4, we looked at the share of the gains from the local processes. We see that potato producers got 33-49% and an average of 40% of the local market price. For the export chain, the producers got on average about 30%. In essence, the percentage potato producers are getting is not low as compared to other chains. But the

denominator for export chains is higher given that export markets attract better prices than the local market. If we assume that local prices of products that are largely consumed in the local market are about 75% of those obtained for export chains (paid in the local market by exporters), the 40% obtained by potato producers is similar to the 30% obtained by the three export chains in Table 1. The amount retained by potato producers can in this instance be considered fair.

But more important is the issue of the investment that other players are making to get the 60-70% margin. For the export chains, there are a lot of investment in governance, packaging and marketing. Packaging costs for the canned peach exports from South Africa constitute 20% of the local costs while it is 25% for the fresh vegetables from Kenya. Kenyan potatoes destined for the local market are packaged very simply in a jute bag that makes up 1-3% of the market price. For the export chains, the local buying agents have to carry out executive governance assisting producers deliver the required quality that would meet the legislated standards in the exports market. This calls for more work by the buyer unlike the case of the buyers of potatoes in Kenya. Basically, they sell ungraded and undifferentiated products where there has been little value addition. There is therefore room to whittle down the trade margins earned by potato traders in Kenya.

CONCLUSIONS

The potato value chain is buyer driven and as expected, the buyer is well compensated for their coordination and control role. Price obtained by farmers is influenced by the market channel, the production geographical region and the size of land cultivated with potatoes and the yields obtained. These last two factors are a proxy for wealth. The farmers who applied the best methods of farming obtained better yields. They are likely to be farming on bigger tracts of land and may perhaps invest more in marketing such that they end up obtaining better prices.

Farmers who sold through other channels besides middlemen obtain better prices. This calls for market upgrade for those who want to obtain better prices. But the smallholder farmers are ill equipped to upgrade their markets. They rely on the middlemen who meet them at the farm gate. Unless the farmer organizations are able to play a facilitative role thereby shortening the very long chain, smallholders will remain captive to the middlemen who have entrenched themselves in the potato trade. Organization of smallholders can enable them fight buyer power as they engage as more significant players who also provide the requisite economies of scale to the buyers.

Though the proportion that farmers retain of the final consumer price approximate what is obtained by producers for agricultural export chains as per Table 5, the investments made by exporter and the attendant risks justify their big margins. For the potato middlemen, the investments are limited to basic items like packaging bags and transport. They do not invest in downstream activities like providing executive governance for farmers to be able to comply with final market requirement. The middlemen do not encourage value addition and most potatoes are sold straight from the farm without even basic post-harvest activities like sorting, cleaning and grading. The governance relationship between the farmers and the buyers is akin to a captive governance structures yet it lacks the key ingredient typical of such a structure where lead firm invest in enhancing the capacity of the small firm to supply solely to themselves. The switching costs are high as the lead firm support will be lost on switching and the processes are too complex for the small firm to handle alone. The suppliers therefore remain captive. For the potato, there is no good rationale for the farmers to remain captive to the middlemen.

The use of extended bag will continue as long as the buyers continue to control the potato value chain. The extended bag allows the buyers to get a bigger quantity for the price of an ordinary bag. The legislation of 110kg bags does not help in the absence of weighing facilities. If weighing facilities were available at the collection point, then buyers may not take advantage of the extended bag as product would be priced for its weight. The government has failed the farmers by not enforcing the marketing standards it has legislated. There is need for an agency that will be able to even the playing ground by carrying judicial governance to rid the market of the extended bag. The Ministry of Agriculture could perhaps take this role as opposed to the current situation where they have been involved in the legislation and continually provide executive governance. Currently, the Ministry of Agriculture laments alongside the farmer at the failure to implement the rules. The Ministry of Local Government which is supposed to enforce the standards at the market place is more concerned about the levies that traders pay and has few incentives to enforce the rules. The Ministry of Agriculture would not need to go to the marketplace which is handled by the Local Government ministry but can monitor enforcement of the rules at the farm gate which is in their domain.

REFERENCES

Crissman C, Crissman L, Carli C (1993). Seed Potato Systems in Kenya: A case study. International Potato Center (CIP).

Durr G, Lorenzl G (1980). Potato Production and Utilization in Kenya. Centro Internacional De La Papa Food and Agriculture Organization Statistics (2010). http://faostat.fao.org FAO Statistics Division. Accessed April 2012.

Gereffi G, Humphrey J, Sturgeon T (2005).The Governance of Global value Chains. Review of International Political Economy 12:1 February 2005: 78-104.

Horton D (2008). Facilitating pro-poor market chain innovation.CIP Working Paper No. 2008-1.

Humphrey J, Schmitz H (2001). Governance in Global Value Chains. Institute of Development Studies. IDS Bulletin 32.3, 2001.

Kaguongo W, Gildemacher P, Demo P, Wagoire W, Kinyae P, Andrade J, Forbes G, Fuglie K, Thiele G (2008). Farmer practices and adoption of improved potato varieties in Kenya and Uganda. CIP Working Paper No. 2008-5.

Kaplinky R (2004). Spreading the Gains from Globalization. Problems of Economic Transitions, vol. 47, no. 2, June 2004, pp. 74-115.

Kenya National Potato Farmers Association (2010). Monitoring and evaluation of implementation of legal notice no. 112, the Local Government Adoptive By-laws on agricultural produce standard weight of packages.

Ministry of Agriculture (2007). Annual report for 2007.

Ministry of Agriculture (2005). Legal notice No. 44 of 2005 (GOK, 2005).

Ministry of Local Government (2008). Legal Notice no. 113.

Republic of Kenya (2007). Kenya Integrated Household Budget Survey 2005/6. Central Bureau of Statistics, Ministry of Planning and National Development .

Republic of Kenya (2011). Economic Survey 2011.

Ruben R, Boekel M, Tilburg A, Trienekens J (2007). Governance for Quality in Tropical Food Chains. The Netherlands: Wageningen Academic Publishers pp. 309.

Trienekens J (2011). Agricultural Value Chains in Developing Countries: A Framework for Analysis. International Food and Agribusiness Management Review 14: 2.

Trienekens J, van Dijk MP (2012). Upgrading of Value Chains in Developing Countries, in: Global Value Chains, Linking Local Producers from Developing Countries to International Markets, pp 237 to 250.

Vorley Bill (2001). The Chains of Agriculture: Sustainability and the Restructuring of Agri-food Markets. International Institute for Environment and Development (IIED) in collaboration with the Regional and International Networking Group (RING).

Wang'ombe JG (2013).Realizing the potential of potato in Kenya, The role of upgrading and governance in the potato value chain. Maastricht: MSM DBA.

Wang'ombe JG, van Dijk MP (2013). Low potato yields in Kenya: do conventional input innovations account for the yields disparity? In: Journal: Agriculture and Food Security, 2: 14, open access DOI 10.1186/2048-7010-2-14MS: 1204114893867422.

8

Livestock Marketing Performance Evaluation in Afar Region, Ethiopia

Tesfaye Berihun

College of Business and Economics, Hawassa University, Hawassa, Ethiopia
E-mail: btesfayenine@gmail.com

This research evaluates the performance of livestock marketing in Afar region, an area characterized by recurrent drought, inadequate basic livestock market infrastructure and accessibility. In addition to secondary data, survey data was collected from 120 randomly selected traders together with intensive price monitoring in seven markets over a period of two months. Markets structures were evaluated using market concentration ratio and; the price co movement, determinants of price and efficiency of the market was also investigated using statistical analysis of multivariate correlation, regression model (ANOVA) and marketing margin respectively. The result indicated that Ayssita, Chifra and Sabure markets had oligopsony market structure with higher wholesale buyers' concentration; and most of the markets were also inefficient with higher marketing margins except for Yallo market. The regression model showed that livestock price was significantly influenced by markets, breeds, gender, age group, and grades of animals and the multivariate correlation result showed, markets were not integrated at all levels. Policies directed to livestock development projects, mainly improving marketing facilities can change the situation there.

Key Words: Livestock marketing, Marketing margins, Market structure, Market integration, Afar region

INTRODUCTION

Ethiopia has the largest livestock population in Africa (Helina and Schmidt, 2012). The livestock sub-sector contributes an estimated 12% to total gross domestic product (GDP) and over 45% to agricultural GDP. The pastoral livestock population accounts for an estimated 40% of the total livestock population of the country (Philimonetal, 2016; Gbremariam and Yemiru, 2015) but when it is compared with the large potential, the existing income generating capacity of livestock is not encouraging (Temesgen et al.2015). Pastoralists in Afar region had 1,990,850 cattle (83.8% share of those animals in the Region in 2003, 2,303,250 sheep 3,960,510 goats, 759,750 camels (CSA, 2003). Central Statistics Agency also estimated in 2005 that farmers in Afar Region had a total of 327,370 cattle, 196,390 sheep, 483,780 goats, 99,830 camels.

Afar is one of the nine ethnic divisions of Ethiopia, and contains the homeland of the Afar people located in north east Ethiopia stretched from the north Danakil depression to south lowland awash valley sharing international boundaries with Eritrea and Djibouti. The region land about 96,707 square kilometers is structured into five zones and 29 woredas (political administration divisions) (Piguet, 2002)

Afar Region is one of the poorest and least developed Regions of Ethiopia. The climate of the region is arid and semi-arid with low and erratic rainfall, bi-modal throughout the region with a mean annual rainfall below 500 mm in the semi-arid western escarpments decreasing to 150 mm in the arid zones to the east (Joanne et al., 2005). The people widely inhabit dry, harsh environment where the temperature sometimes rises to 48 ^{0}c in the hottest months. The northern part of the region is the driest and hottest part of the region including Dalol depression, one of the lowest and hottest places on earth 116 m below sea level. The people rely on livestock for survival; more than 90 per cent of the population depends on cattle, sheep, goats and camels as a source of food and cash (Gbremariam and Yemiru, 2015). They move from one area to another depending on the availability of seasonal water and grazing land. Sometimes in different parts of the region conflict forces pastoralists to change their usual migration patterns and most importantly were denied access to either traditional water points and wells or grazing areas or both together. On top of this rather complex and confuse conflict situation, drought due to lack of rain in the region forces the pastoralists to migrate in neighboring regions (Piguet, 2002).

Preliminary studies and reviews (Belachew and Jemberu 2002; Aklilu 2002) indicate the structure and performance of live animal market both in export and domestic markets of the country are poor. The wide-ranging problems of Underdevelopment and lack of market-oriented production, lack of adequate information on livestock resources, inadequate permanent animal route and other facilities like water and holding grounds, lack or non-provision of transport, ineffective and inadequate infrastructural and institutional set-ups, prevalence of diseases, illegal trade and inadequate market information (internal and external) negatively affects the overall Ethiopian livestock trade. Despite these are some of the major problems of livestock marketing outlined in the nation at large, specific evaluation of the region's livestock marketing performance with intensive market throughput and price from the major markets is not sufficiently available so far. Peripheral areas of the country have an absurd gap in livestock market information (FAO, 1999) and Afar region in particular have suffered lack of market information and support on access and market infrastructure (Piguet, 2002). In absence of data on the magnitude and seasonality of supply and prices analysis; and without market evaluation livestock market development projects and institutional supports will not grant the desired success. So, the wide information gap in marketing system and market integration through alternative channels of the region needs to be narrowed as the main component of livestock market development effort. Therefore, this study evaluates the livestock marketing performance (live animal of cattle, camel, sheep and goat) of the Afar region and tries to justify whether there are reflective and actual problems prevailing in the region and likely to cause poor performance of the sector in the nation as it is reviewed in the widest sense. And the study will also give a clue to the comparative advantage of livestock marketing proposal and strategic development options in assessing and determining the degree of market efficiency in terms of marketing margins and spatial price correlation of markets. In broad-spectrum, this study on the above issue will also help policy makers to fine-tune policy decision and supportive intervention of livestock marketing in the pastoralist context.

METHODOLOGY

According to the Regional Agriculture Pastoral and Rural Development Bureau at preliminary survey, seven markets (Asayita, Chifra, Yallo, Sabure, Abala, Gewane, and Amibera/werer) were recognized as the main markets specialized in all specious (sheep, goat, cattle and camel) in the region. These markets are also known by high volume of weakly animal supply and their linkage to terminal and export markets. Accordingly, all the seven markets are purposely selected for both traders' survey and market price monitoring. A total of one hundred and twenty sample traders have been randomly selected and interviewed from the list of traders operating in the seven markets. Quantitative survey through cross-sectional market price monitoring system were also collected from these markets by direct observation and discussion with key informants at a time in two months on the basis of market days (short term intensive weakly price monitoring). Other market data from secondary sources (Collected by GL-CRSP Livestock Information Network Knowledge System) was also used. To capture seasonal variation and changes due to religious festivals, time-serious data were used from secondary sources.

Analytical tools

In order to evaluate the livestock marketing performance of the region market concentration ratio, marketing margin analysis, Analysis of Variance and multivariate correlation coefficient were` used.

Marketing Margin

In measuring the marketing margin cross-sectional margin analysis method were used in a stepwise analysis through:
- Indicating the market structure and linkages
- Making out the main physical and facilitating operations performed between the different outlets specified
- Estimating the cost of operations including hidden costs performed within each outlet

FAO, (1999) Reference standards can be used to set up a point at or beyond which performance is judged to be "satisfactory" or "unsatisfactory". Market margins of more than 15%, for example, could be considered unacceptable. Efficiency in performance of marketing functions is not in all cases equated with small marketing margins. Similarly, large margins are not necessarily a firm indication of inefficiency or excess profit by traders. Marketing margins and costs can only be meaningfully discussed in relation to the services and functions which are provided.

Market Concentration

FAO, (1999) Market structure is a measure of market evaluation which can complement the market margin analysis. Market structure refers to the characteristic of market participants which technically influences the market behavior in determining competition and pricing system. The basic indicators of a particular market structure are the number of participants and their size distribution in the market that entail the degree of market concentration. Market concentration ratio refers the total numbers of buyers/sellers and their size distribution to have a share in a market.

Analysis of Variance (ANOVA)

Kother (2004), through ANOVA technique it is possible to investigate any number of factors which are hypothesized or said to influence the dependent variable and one may as well investigate the differences amongst various categories within each of these factors. So in this research ANOVA model is used to assess the statistical significance of the relationship between Price of animals and qualitative regressors that are assumed to be factors for livestock price variation (Market, Breed, Gender, Age and Grade of animals) and it is also used to compare the differences in the mean values of the factor groups hypothesizing significant mean price difference.

Multivariate Correlation Coefficient Analysis

Based on the short term price monitoring data, a spatial price correlation between animals and grades in same market and market integration between same grades of animals in different markets are analyzed using multivariate correlation coefficient analysis. It is a widely-used evaluation method which looks at the spatial correlation of markets through time. The assumption is that if market prices in different regions move together, then the overall market is operating effectively, in that supply is being distributed regionally in a way which meets local demand.

Variables

Dependent variable:

Price of Livestock

Independent variables:

Age of animals: It is used as a discrete dummy variable having three classes of age groups such as immature, young, and matured age classes. (Conventional by traders)

Sex of animals: Discrete dummy variables represented in the model by 1 and 0 for sexes male and female.

Grades of animals: There are four grades of animals that represents; grade1 represents large mature animals for export quality. Grade 2 represents medium size export quality. Grade 3 represents young animals having good body condition used by export abattoirs. Grade 4 represents animals greater than immature and having average body condition mainly used for local consumption.

Market places: Asyita, Chifra, Yallo, Abaala and Werer markets were also taken as a discrete dummy variable.

Breed of cattle: All the cattle breeds in the region (Arsi, Boran, Danakil, Mixed, Raya Azebo and Zebu) were also taken only in Cattle category.

RESULTS AND DISCUSSION

Market Structure

The survey result indicates livestock marketing in the region lacks specialization in transacting the product from producers to consumers. At any markets all pastoralists, local traders, medium and large scale assembler traders, buying agents and consumers participate in the transaction process and then animals are channeled from producer to consumers and to other assembler traders through them. Pricing system in the markets is in fact a result of market structure that is a high market concentration results inefficient pricing system and the inverse is true, that lower concentration results in market efficiency assuming that competition will clean-up any of unnecessary margins in competitive market.

The market concentration ratio of the four selected markets (Assayta, Chifra, Yallo and Sabure) is computed using the most commonly measuring system known as market concentration index of a four-firm concentration ratio [1](CR4). This ratio implies, if the market share of the largest four firms is less than or equal to 33% generally indicates a competitive market structure, while a concentration ratio which is 33% to 50% implies weak oligopsony market and the market share of the largest four firms above 50% indicate strong oligopsony market structures Kohls and Uhl (1985).

The numbers of traders who are assembling animals from each market vary time to time depending on drought conditions. However, there are a number of traders who collect animals from each market regularly. As it is shown in Table 1, the concentration ratio result implies the existence of strong oligopsony market structure in Ayssita market especially in shoat (sheep and goat) and cattle marketing with a CR4 75% and 52% respectively. Weak oligopsony market structure is observed in Chifra and Sabure markets with CR4 between 33% and 50% excepta competitive shoat markets in Chifra CR4 30%. Yallo market is the only competitive market having a lower four

Table 1: Four firms market concentration ratio at selected markets

Markets	Animals	CR4 (%)
Ayssita	Cattle	52
	Shoat	75
	Camel	44
Chifra	Cattle	38
	Shoat	30
	Camel	41
Yallo	Cattle	19
	Shoat	12
	Camel	26
Sabure	Cattle	36
	Shoat	47
	Camel	38

Source: Author's computation

Table 2: Cattle and Shoat (Sheep and Goat) marketing costs and margins at regional and terminal markets

Description	Cost, Price and Margins (USD. per head)					
	Cattle	Shoat	Cattle	Shoat	Cattle	Shoat
Mean purchase price from producers	119.8	16.5	120.7	17.4	103.0	15.5
Market costs up to regional markets	Assayta		Yallo		Sabure	
- trekking	0.9	0.1	0.8	0.2	0.7	0.1
- watering fee	0.3	0.1	0.2	0.1	0.1	0.1
- food and lodging	0.4	0.2	0.3	0.3	0.2	0.2
- loss - trading	0.2	-	-	0.1	0.2	0.1
- others	0.2	-	0.2	-	-	-
Total	2.0	0.4	1.5	0.7	1.2	0.5
Mean sales price at regional markets	130.9	19.9	130.0	21.6	114.1	18.7
Trader's mean gross margin at regional market	8.5	3.0	7.8	3.5	9.9	2.7
Mean purchase price at regional markets	130.9	19.9	130.0	21.6	114.1	18.7
Marketing costs up to terminal markets	Djibouti		Mekelle		Nazret	
- County Council fees	0.3	0.2	0.3	0.2	0.3	0.2
- trekking fee	3.3	-	1.5	-	0.5	-
- watering fee	0.4	0.2	0.2	0.3	0.2	0.1
- food & lodging	0.6	0.3	0.2	0.4	0.4	0.2
- transport	7.0	0.8	6.5	1.0	4.0	0.6
- miscellaneous costs	0.2	0.2	0.2	0.2	0.1	0.2
- loss - trading	0.2	-	0.1	-	-	-
Total	12.0	1.7	9.0	2.1	5.5	1.3
Mean sales price at terminal markets	153.3	26.3	148.2	27.2	129.8	23.8
Trader's mean gross margin at terminal markets	11.0	4.3	9.2	3.5	10.2	3.8
Total Gross Marketing Margin (TGMM)	21.85%	37.26%	18.56%	36%	20.64%	34.8

Source: Author's computation

firm's concentration in all Shoat, Cattle and Camel markets. This market is mainly influenced by large traders from Tigray region. Relatively large volumes of camel are traded from this market that is going to be exported to Sudan through Humera.

Marketing Margin

Prices of uniform products at each market stage cross-sectional at one point in time across a variety of market agents are used to determine livestock marketing margins. In computing the marketing margin at selected level of marketing channels both direct and indirect cost estimation of physical and facilitating function of marketing are made

in to use. For this study markets linkages are disaggregated among the main channels of market chains in the region. These markets are Ayssita, Yallo, and Sabure vertically linked with [2]terminal markets of Djibouti, Mekelle, and Nazret respectively. Other markets have direct relation in feeding the three regional markets in the chain towards the final markets.

In every stages of the marketing chains the direct cost of transportation and assembling are used with no complexity. While the indirect cost of transportation like death and loss of weight are calculated in adding up the weighted average loss on the remaining traded animals. Due to the traditional marketing system at each market

Table 3: Regression Result (ANOVA)

	Rsquare	F Ratio	Prob> F		Rsquare	F Ratio	Prob> F
Model	0.682029	289.5667	<.0001				
Sheep price				**Camel price**			
Market	0.289599	123.4177	<.0001	Market	0.122644	27.5384	<.0001
Age group	0.238236	189.6780	<.0001	Age group	0.634628	514.1337	<.0001
Gender	0.017587	21.7324	<.0001	Gender	0.01108	6.6438	0.0102
Grade	0.143898	67.9066	<.0001	Grade	0.063705	20.1398	<.0001
Goat price				**Cattle price**			
Market	0.06745	25.1524	<.0001	Market	0.083819	27.9266	<.0001
Age group	0.300282	298.9004	<.0001	Breed	0.095449	25.7472	<.0001
Gender	0.012819	18.1016	<.0001	Age group	0.328595	299.2763	<.0001
Grade	0.183246	104.1028	<.0001	Gender	0.001323	1.6215	0.2031
				Grade	0.087868	58.9072	<.0001

Source: Author's computation

Table 4: ANOVA- Mean price of Animals

	Goat		Sheep		Cattle		Camel	
	Number	Mean(Std.error)	Number	Mean(Std.error)	Number	Mean(Std.error)	Number	Mean(Std.error)
Market								
Abaala	355	23.27(0.44)	309	30.43(0.50)	519	140.68(2.40)	231	254.02(4.79)
Ayssaita	386	18.12(0.42)	302	17.46(0.50)	282	122.40(3.25)	118	194.83(6.70)
Chifra	240	18.26(0.53)	218	18.72(0.59)	126	114.06(4.86)	113	190.15(6.85)
Werer	237	17.73(0.54)	210	16.46(0.61)	151	89.72(4.44)		
Yallo	178	19.24(0.62)	177	18.30(0.66)	148	123.38(4.49)	133	223.89(6.31)
Age group								
Immature	332	11.85(0.39)	248	11.08(0.58)	382	77.78(2.39)	183	138.00(3.47)
Mature	596	23.85(0.29)	548	24.38(0.39)	496	153.20(2.10)	197	291.24(3.34)
Young	468	19.47(0.33)	420	22.25(0.44)	348	137.96(2.50)	215	233.98(3.20)
Gender								
Female	582	18.38(0.35)	467	19.19(0.48)	572	123.16(2.38)	107	240.83(7.46)
Male	814	20.35(0.30)	749	22.02(0.38)	654	127.31(2.22)	488	219.60(3.50)
Grade								
Grade 1	8	48.12(2.74)	22	43.81(2.05)				
Grade 2	547	22.61(0.33)	455	23.36(0.45)	208	154.32(3.77)	96	265.55(7.67)
Grade 3	637	18.25(0.31)	543	19.51(0.41)	661	128.05(2.12)	374	219.34(3.89)
Grade 4	204	14.13(0.54)	196	16.65(0.69)	357	103.57(2.88)	125	203.26(6.72)
Breed								
Arsi					263	126.46(3.34)		
Boran					4	85.50(27.13)		
Danakil					601	113.00(2.21)		
Mixed					9	156.0(18.08)		
Raya Azebo					236	132.21(3.53)		
Zebu					113	173.40(5.10)		

Source: Author's computation

yards in the region, none of the costs of [3]facilitating functions; finance, risk bearing, and information are came out as direct costs and as a result these are not included in the cost estimation directly.

Generally, the lower the marketing margin the more markets tend to be efficient. But to determine the overall market efficiency, operational and price efficiency should be examined from the producers and consumers point of view. Compatible cost increases with the demands of consumers as a result of improved service may cause higher marketing margin, but this doesn't imply the market is inefficient. The market will remain inefficient only when

the cost becomes higher without additional service or value. In this respect the retail price change at all links of the selected market channels in the region is associated with traders margin and transportation cost. These costs in fact have no more value addition other than the place value. So, since there is no extra service or value addition for the change in retail price, the marketing margin can show efficiency of markets within certain economic context. As a result, 18.56% TGMM of cattle marketing makes the channel from Yallo to Mekelle relatively efficient. As shown in Table 2, cattle marketing in all the channels relatively perform well with lower TGMM (21.85% Ayssita to Djibouti, 18.56% Yallo to Mekelle and 20.64%

Table 5: Pairwise Correlations result of the same types of animals in different markets

Pairwise Correlations result of Sheep price				Pairwise Correlations result of Goat price			
Variable	by Variable	Correlation	SignifProb	Variable	by Variable	Correlation	SignifProb
Ayssaita	Chifra	0.3552	0.0008	Ayssita	Abaala	0.5666	0.0000
Abaala	Chifra	0.0362	0.7406	Chifra	Abaala	0.3191	0.0027
Abaala	Ayssaita	0.2938	0.0060	Chifra	Ayssita	0.2398	0.0262
Werer	Chifra	-0.0524	0.6316	Werer	Abaala	-0.1749	0.1072
Werer	Ayssaita	0.0797	0.4660	Werer	Ayssita	-0.3614	0.0006
Werer	Abaala	0.0218	0.8422	Werer	Chifra	0.2089	0.0535
Yallo	Chifra	0.0584	0.5930	Yallo	Abaala	0.0991	0.3638
Yallo	Ayssaita	0.0639	0.5590	Yallo	Ayssita	0.1763	0.1044
Yallo	Abaala	-0.0818	0.4541	Yallo	Chifra	0.1814	0.0945
Yallo	Werer	0.0920	0.3995	Yallo	Werer	-0.1920	0.0766
Pairwise Correlations result of Cattle price				Pairwise Correlations result of Camel price			
Variable	by Variable	Correlation	Signif Prob.	Variable	by Variable	Correlation	Signif Prob
Ayssita	Abaala	0.0940	0.4710	Ayssita	Abaala	0.3031	0.0259
Chifra	Abaala	0.0638	0.6251	Chifra	Abaala	0.0436	0.7545
Chifra	Ayssita	0.5947	0.0000	Chifra	Ayssita	-0.1738	0.2089
Were	Abaala	0.0646	0.6211	Yallo	Abaala	-0.1617	0.2429
Were	Ayssita	0.4832	0.0001	Yallo	Ayssita	-0.0508	0.7151
Were	Chifra	0.3684	0.0035	Yallo	Chifra	0.1519	0.2727
Yallo	Abaala	0.0010	0.9941				
Yallo	Ayssita	0.1053	0.4193				
Yallo	Chifra	0.2097	0.1048				
Yallo	Were	-0.1147	0.3789				

Source: Author's computation

Sabure to Nazret). Shoat marketing in all channels seemed to be inefficient with higher marketing margin. Coordinated supply chains along the market channels by market actors can reduce transaction costs.

Price Analysis

Price as a main continuous dependent variable was regressed with discrete variables that are livestock breed, age group, gender, market places and grade of animals. The whole regression model shows significant relation between price and all the regressors, except gender in cattle price.

To identify the mean price difference within the groups of each factor for each species one way analysis of variance was conducted.

Goat price

One way analysis of variance of goat price shows that there is a significant mean price variation in all groups. As it is shown in Table 4, Abaala market has significantly high mean price difference from other markets and the rest of four markets mean goat price is not considerably different. The result also shows there is significant mean price difference among all age groups with high price of mature ages and relatively low price of immature ages. Regarding to gender, male goats have significantly high price than female goats. Grade one goats also have highest price

among all groups and grade four goats have much lower price.

Sheep price

The analysis of variance result of sheep price is the same as goat price result. Only Abaala market has significantly high mean price variation from the other four markets. There is no significant mean price variation among the rest four markets. But other factors like age, gender, and grade of animals have significantly different mean price variation. Male sheep have higher price than female and the higher the age of sheep's (mature, immature, and young) and the superior the grades of animals the price will get higher and higher.

Cattle price

The most traded cattle breed in all parts of the region is Danakil breed. And Raya Azabo breed is common in the northern part of the region. The statistical analysis result indicates that price variation exists among markets. As displayed in Table 4, Abaala markets average cattle price is significantly different from other markets having higher mean price of 140.6USD and Werer market as well have a minimum mean price 89.7USD which is considerably different from other markets. The rest markets of Chifra, Yallo, and Ayssita have no significant cattle price variation. Price difference among cattle markets may be resulted from the domination of typical cattle breed in the market. Significant mean price variations among all breeds are observed. Zebu breed has highest mean price 173.4USD followed by Mixed breed 156 USD. While Borena and Danakil breeds have lower mean price of 85.5 and 113 respectively. There is also significant price difference among all grade and age groups. But the price difference of male and female cattle in immature age group is not remarkable.

Camel price

Considerable camel price difference is found between the markets. Abaala and Yallo have higher mean price of 254 USD and 223.9 USD respectively whereas in Ayssita and Chifra market camel price variation is not significant with lower average price of 194.8 USD, and 190.1 USD respectively (Table 4). Matured age Camels have much higher price and it decreases as the age decrease in the range. There is also significant price change between male and female camel.

Abaala market is found in the driest and hottest part of the northern region where there is no as much local livestock markets as in the rest of the region. So absence of alternative markets in the nearby locality could be the reason for high price of all specious in Abaala market.

Impact of seasonality on livestock price

Livestock of Matured age group (the most traded age groups with large body condition for both export and local consumption) is selected for this seasonality analysis. Seasonality in general refers different climatic conditions within a year. The largest part of Afar region remains very hot all the year round and other parts which are adjacent to the highlands of the country experiences the four seasons. The general pattern of price depends on the availability of pasture and water over the year. During the dry season (mid-December to mid-march) general livestock price tends to decline due to the drought that causes pastoralists to sale their animals. Depending on the rain condition, if it comes early, price of livestock will go up between mid-March to mid-June. It is noticeable that festivals and demand in domestic and export markets increase price along the year.

When we look at Camel price, it is high in January and becomes low to June and starts to rise to August. And there is also price increase in September and remains slightly lower until December. Cattle price also starts to decline from early February to May and rise the month onwards till it reaches to the peak in August. Then the price remains the same at average rate to December.

Goat and Sheep prices trend is almost the same. Both Goat and Sheep price starts to turn down at declining rate from January to March and becomes high rapidly at April, then goes down to June and remains high again between August to October, and it begins to decline in December.

This analysis lacks dependability to conclude the general trend of livestock price other than the year under consideration.

Price co-movement and market integration

Market performance can be analyzed using the indirect market efficiency measures of spatial price behavior that is price co-movement. Price co-movement indicates degree of price integration (how linked markets are across different channels). This can be studied using multivariate correlation coefficient. Price at a given market is assumed to transmit or co- moves to other markets through arbitrage in a competitive and efficient transaction system. Here we see livestock price integration in two ways; different specious and grades of animals in the same market and, the same grades of animals in different markets. If the Multivariate correlation coefficient exceed +0.6 the market is known to be integrated (Blyn, 1973)

Multivariate correlation analysis of the same types of animals in different markets

Positive multivariate correlation coefficients of particular grades of animals in a number of different markets indicate there is potential integration and the markets are likely to function well. Animals of matured age group at each species are used to the multivariate correlation coefficient analysis of the five markets (Abaala, Ayssita, Chifra, Werer, and Yallo).

Sheep

Significant price correlation coefficient is observed only between Ayssita and Chifra markets (p = 0.0008) and Abaala and Ayssita markets (p = 0.006). The result generally indicates, there is no significant price correlation among the majority of markets except the moderate price co-movement observed between Ayssita and Chifra markets and Abaala and Ayssita markets with positive correlation coefficient 0.355 and 0.294 respectively (Table 5).

Goat

Relatively significant positive price correlation is observed in goat market between Ayssita and Aballa, Chifra and Abaala, and Chifra and Ayssita markets with correlation coefficient of 0.566, 0.319, and 0.239 respectively (Table 5). Unexpectedly, a significant negative price correlation is observed between Werer and Ayssita markets. Werer and Ayssita markets are geographically located in a distant from each other and served as sources of livestock supply to the border Djibouti market and Nazret market respectively. The high cost of exporters and abattoirs that collects Goat at Werer market may cut the price down there, and possibly the demand against the supply of

Ethiopian exporters at border market can pick up the price at Ayssita market. This is a plain assumption behind the negative correlation between them.

Cattle

Significant positive cattle price correlation is observed between Chifra and Ayssita, Werer and Ayssita, and Werer and Chifra markets with correlation coefficients 0.595, 0.483, and 0.368 respectively (Table 5)but the markets are not yet said to be well integrated.

Camel

Significant positive Camel price correlation is found only between Ayssita and Abaala markets with correlation coefficient of 0.30. The significant positive correlation coefficient shows that there is a price co-movement but it is not enough to say the markets are well integrated

Generally the degrees of price co-movement between markets are low. In large numbers of market pairs the animals' price correlation coefficient is close to zero. That implies no correlation and co-integration is found at all between markets. But in relative sense Ayssita market is better correlated with some other markets followed by Chifra market. Inadequate marketing services such as lack of timely market information and inadequate market infrastructures causes lagged response of price to the demand and these can be the main reasons to lack of price co-movement.

Multivariate Correlation analysis of different types of animals at same market

Positive correlation coefficient of different specious or grades of animals in the same market indicates potential substitution for one another. Strong price co-movement is observed between different grades of same animals in the major markets. There is a strong price correlation between Heifer & Fattened ox, Steer & Fattened ox, and Steer & Heifer with correlation coefficients of 0.833, 0.854, and 0.665 respectively. And a strong price co-movement is observed between Goat (buck) and Ram (sheep) in Ayssita market. No price co-movement is observed between large stocks and small stocks; this can indicate lack of substitution between them.

CONCLUSION

This study evaluates marketing performance in view of market structure, marketing margin and market integration. Organizational characteristics of a market and its influence on the nature of competition and pricing is mainly explained by market structure; the result indicates that most of the regions' markets are dominated by oligopsony market structure with a higher market

concentration ratio of wholesale traders, as a result the largest market share go away with these groups and they also have the power to influence price. Yallo market is the only competitive market having a lower four firm's concentration in all Shoat, Cattle and Camel markets. The price difference from each local and regional markets to final markets in all main livestock market channels are evaluated and the result indicates Yallo market to Mekelle market is relatively efficient in all animals' marketing with lower marketing margin and cattle marketing in particular is relatively efficient in all market channels compared to shoat and camel marketing. Improving marketing facilities can reduce direct costs and indirect costs of transportation like weight loss and death. The regression analysis shows livestock price is influenced by livestock breed, age group, gender, market places and grades of animals. The one-Way Analysis of Variance also reveals there is significant mean price variation among each price determining factor groups. Availability of pasture and water largely affects the general pattern of price over the year. During the dry season (mid-December to mid-march) general livestock price tends to decline due to the drought that causes pastoralists to sale their animals and if rain comes early price will go up between mid-March to mid-June. In integrated markets, price shocks from one geographic market are transmitted to other markets through the trading system and the supply of animals adjusts spatially to meet demands but in this study result shows the degrees of price co-movement between markets in general are low and that implies markets are not integrated. Relatively Ayssita market has better price co-movement with other markets. Strong price co-movement is observed between different grades of same animals in the major markets, and also strong price co-movement is observed between Goat (buck) and Ram (sheep) in Ayssita market that indicate potential substitution between them. Weak information and infrastructure and high transport and other marketing costs can be the reason for the poor market integration so working on marketing facilities can improve the situation in the regions livestock marketing system

REFERENCE

ACDI/VOCA (2010). Impact Assessment of the ACDI/VOCA Livestock Markets in Pastoralist Areas of Ethiopia. PLI Policy Project Tufts University November 2010

Akililu, Y. (2002). An Audit of the Livestock Marketing Status in Ethiopia, Kenya, and Sudan. Volume 1. Nairobi, Kenya.

Belachew Hurissa and Jemberu Eshetu. (2002). Challenges and opportunities of livestock trade in Ethiopia. Paper presented at the 10th annual conference of Ethiopian Society of Animal Production (ESAP), Addis Ababa, Ethiopia, 22–24 August 2002. ESAP, Addis Ababa, Ethiopia.

Blyn, G. (1973). Price series correlation as a measure of market integration. Indian Journal of Agricultural Economics 28.2; 56-59, New Delhi: Indian Council of Agricultural Research

CSA (Central Statistical Authority), Report on Monthly and Annual Producers' Prices of Agricultural Products in Rural Areas by Killil and Zone, (2003, 2005), Addis Ababa.

FAO (1999) Food and Agriculture Organization of the United Nations FAOSTAT 1999 http://apps.fao.org/cgi-bin/nph.db.pl.

GL-CRSP (2007) Livestock Information Network Knowledge System, Market data Chart.Version: LINKSV3.042409_testBuild(http://www.lmistz.net/Pages/Public/chart.aspx)

Helina T. and E, Schmidt (2012). Spatial Analysis of Livestock Production Patterns in Ethiopia. ESSP II Working Paper 44. Addis Ababa, Ethiopia: International Food Policy Research Institute/ Ethiopia Strategy Support Program II, Ethiopia

Joanne P, Asnake A, Kassaye H (2005). Livelihood emergency assessment in Afar region. OXFAM International. Addis Ababa, Ethiopia. P 44.

Kohls R L and Uhl J W. (1985). Marketing of agricultural products. Sixth edition, Macmillan, New York, USA.624 pp.

Kother, C.R (2004). Research methodology, methods and techniques. New Age International (P) Ltd., Publishers, New Delhi, India. P 257

Philimon T., D. Kidanie, T. Endeshaw, K. Ashebir, T.Abebe, G.Weldegebrial, S.Workinesh and T. Woldegebriel (2016) Study on cattle management and marketing practices in Afar region. Int. J. Livestock Production. Vol. 7(8), pp. 55-65

Piguet F.(2002),Afar: insecurity and delayed rains threaten livestock and people- Assessment Mission: 29 May – 8 June 2002UN-Emergencies Unit for Ethiopia, Addis Ababa, Ethiopia.

Temesgen G, Aleme A, Mulata H (2015). Climate change and livestock production in Ethiopia. Academia J. Environ. Sci. 2(4):059-062.

9

An effect of support price toward the growth rate of sugarcane production: Evidence from Sindh and Punjab provinces of Pakistan

Mansoor Ahmed Koondhar[1*], Abbas Ali Chandio[1], He Ge[1], Mumtaz Ali Joyo[2], Masroor Ali Koondhar[2], Riaz Hussain Jamali[2]

[1]College of Economics and Management, Sichuan Agriculture University, Chengdu, 611130 China.
[2] Department of Agricultural Economics, Sindh Agriculture University Tandojam, Pakistan.

Co-authors   Email: [1]3081336062@qq.com,[1]Hege01@126.com,[2]joyo.mumtaz@gmail.com,[2]masrooralikoondhar@gmail.com,
[2]jamalirh@ahoo.com,

This paper focuses on an effect of support price toward the growth rate of Sugarcane production: Evidence from Sindh and Punjab provinces of Pakistan by using secondary time series data from the period of 1990-91 to 2013-14. Growth rate model and Cobb-Douglas production function was applied to analyze the data. Every year government of Pakistan announced support price for sugarcane crop, the aimed of announcing supporting price to save the sugarcane producers for achieving the target of sugar production. In Punjab province, since 1990-91 to 2013-14 total growth rate of sugarcane in area, production and yield were increased 2.24%, 4.67% and 2.33% respectively. However, in Sindh Province total growth rate was calculated 1.42% for area, 3.35 for production and 1.78% for yield respectively. The results of regression analysis indicate that the both province`s area and support price have significant relationship with production. However its necessary to increase support price if the support price increase than the farmers take keen interest for cultivating more area under sugarcane with use of modern technologies and also increase the applications of inputs, so that the government of Pakistan should increase support price for promoting the sugarcane production both Sindh and Punjab provinces of Pakistan.

Key words: Sugarcane production, support price, growth rate, cobb-Douglas, Sindh and Punjab, Pakistan

INTRODUCTION

Agriculture play a vital role in Pakistan`s economy. This sector provides raw material to agro based industries. The most significant role of agriculture in the economy of Pakistan is to surpluses generate for export to earn foreign exchange. In 2015 its account 20.9 percent in the grass domestic product (GDP) and 43.5% is the source of earning for rural peoples, furthermore agriculture effective supporting sustainable sophisticated economic growth and also a significant impact on reducing of poverty in Pakistan. Agriculture sector has traditionally sustained a satisfactory growth to ensure food security for rapidly growth of population. Yet, the main challenges faced the farmer's low returns of agricultural commodities due to high cost of production. (GOP 2014-15)

*Corresponding Author: Mansoor Ahmed Koondhar, M.Sc. Scholar, College of Economics and Management, Sichuan Agriculture University, Chengdu, 611130 China. Email: 3115059778@qq.com

Sugarcane is the most important cash crop of Pakistan. Sugarcane mostly cultivated for producing sugar, and correlated products, such as Gur, sugar used for human diet, fuel and fiber has been advocated through by products (Deepchand 1986). About 99 percent of sugar extracted from sugarcane in Pakistan for demand at national level (Azam and Mukarram 2010). Sugarcane also have significant relationship with Board industries for making papers. Sugarcane account 3.1% in value addition and 0.6% share in GDP. (GoP 2014-15) sugarcane was cultivated on area of 1141 thousand hectares in the year of 2014-15 it is decreasing 2.6% beside preceding year of 2013-14 the area was cultivated 1173 thousand hectares. The production of sugarcane was recorded 62.7 million tons in 2014-15 it was 7.1% less than last year 2014-15.The factor affection reducing of crop production due to decreasing Area, heavy rainfall flood, climate changes the main effect of reducing production due to announced unwanted policies about supporting price of sugarcane crop.

In Pakistan sugarcane generally cultivated in three provinces Sindh, Punjab and NWFP (Masood and Javed 2004). The highest production of sugarcane was recorded in Punjab 37,704, thousand tons in 2013-14, in Sindh 18,362, thousand tons and NWFP, 5,361, thousand tons, Sindh and NWFP received lower production as compare to Punjab but the yield of Sugarcane/hac highest was recorded in Sindh 61.71 tons/hac, Punjab 57.75 tons/hac and NWFP received lower 45.68 tons/hac.

In Pakistan sugarcane industries is second largest agro based industry after textile industries, its account 3.25 in value addition and 0.6% in GDP (LCCI 2013), There are total 88 Sugar mills, 44 sugar mills in Punjab, 36 in Sindh and 08 in NWFP, rapidly growth of sugar industries also contribute in the growth of economy of Pakistan, sugarcane is one of the biggest revenue source of Pakistan`s government, due to sugarcane fetches billion of rupees in the farm of duties and taxes to government (Adnan Nazir 2013), its employee 1.5 million workers, including; financial experts, management experts, technologist (engineers), skilled, semiskilled and unskilled workers working in sugar industries of Pakistan (LCCI[5]).

The contribution of sugarcane in GDP, employment, foreign earning and rising the income of farmers, however the government of Pakistan announced supporting price of sugarcane to support the farmers for cultivating more area and to use modern technologies with maximum inputs for the achieving of government targets, (Magsi 2012) The main cause of conducting this study was to compare the effect of supporting price on growth rate of sugarcane production in (Sindh and Punjab) Pakistan.

METHODOLOGY

The Study entitled an effect of support price towards the growth rate of sugarcane production: Evidence from Sindh and Punjab provinces of Pakistan, was conduct for 24 years on the basis of secondary time series data, since the period of 1990-91 to 2013-14. The data were collected from Pakistan Sugar mills Association. The data were collected regarding area, production, Yield and supporting price which announced by Pakistani government. Both the micro and macro data were analyzed through use of Excel and SPSS software. Growth rate model and Cobb-Douglas production function were used for the finding of how much growth in sugarcane with time index? Which independents variable impact on sugarcane production? And how supporting price effecting on sugarcane production?

Growth rate Model: secondary time series data were analyzed sugarcane production, area and yield for the estimation of growth rate.

$$g_x = X_T - X_0 \left(\frac{}{X_T} \right)$$

g_x= growth rate
X_T= 1[st]year value of variable X
X_0= 2nd year value of variable X
T = 1st year
0 = next year

Cobb-Douglas production function was specified as under:

$$Y = f(X_1^{\beta 1}, X_2^{\beta 2})$$
$$LnY = \beta o + \beta 1\, Ln\, X1 + \beta 2\, Ln\, X2$$
$Ln\beta 0$= Natural logarithm of sugarcane in (000) tones.
Ln X1= Natural logarithm of sugarcane cropped area in (000) cultivated hectares
Ln X2= Natural logarithm of sugarcane crop supporting Price/40kg.
μ= error term

RESULTS AND DISCUSSION

This study presents growth of sugarcane production, area and yield in Sindh and Punjab provinces, and to estimating relationship of area and support price with production of sugarcane in Sindh and Punjab provinces of Pakistan.

Figure1, 2 and 3 represents sugarcane production area and yield difference in both provinces of Pakistan, Figure No 1 and 2 indicate that the area and production growth with time index in Punjab and Sindh provinces of Pakistan in Punjab are higher than Sindh Province, the higher area under sugarcane cultivation recorded in 2007-08, 872.2 thousand hectares and production was 40,306.0 thousand tons but in 2008-09 the area of sugarcane was cultivated 666.5 thousand hectares

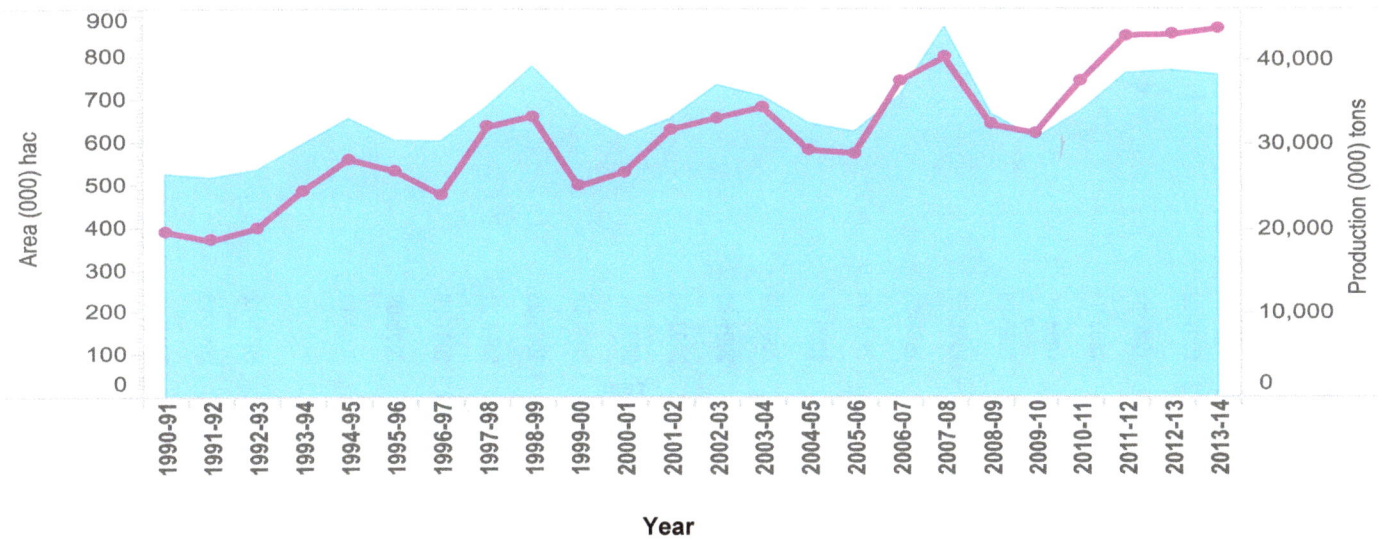

Year
Data Source: Pakistan Sugar Mills Association
Figure 1. Sugarcane Production and Area with time index in Punjab province of Pakistan

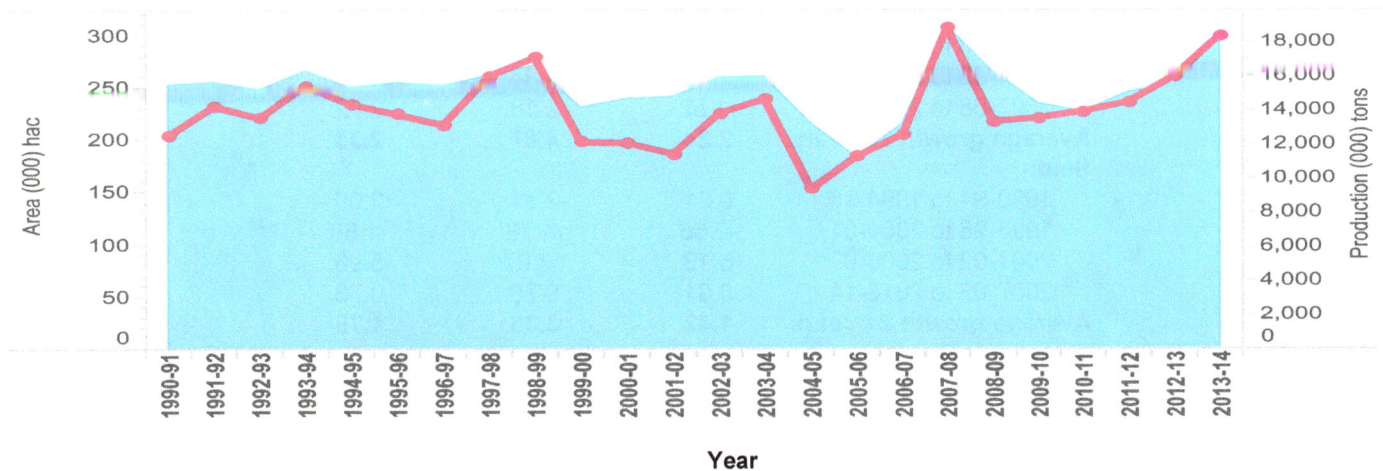

Year
Data Source: Pakistan Sugar Mills Association
Figure 2. Sugarcane Production and Area with time index in Sindh province of Pakistan

and production was 32,294.7 thousand tons. The area was decrease, 23.5% and the production was decrease 19.8% less then area, Since 2008-9 to 2013-14 the area of Punjab sugarcane cultivation increase 13.5% and the production was increased 35.3 % it is double than area. In case of Sindh province the area of sugarcane was cultivated more 308.8 thousand hectares in 2007-08 in same year more production received by growers 18,793.9 thousand tons but the area and production of sugarcane in Sindh province also decreased in 2008-09 till to 263.9 thousand hectares and production was decreased 13,304.3 thousand tones. The decreasing ratio of area in 2008-09 was 14.5% and 29.2% production decrease more than area, since 2008-09 to 2013-14 area was increased 12.7% and the production was increase

38% it is 3re times more than area, and the figure 3 Indicate that the difference of sugarcane yield in Sindh and Punjab provinces with time index. In Sindh province yield of sugarcane always be more than Punjab province the highest yield was recorded in Sindh province 1574 mnds/hac in 1998-99 in same time period the yield of sugarcane was recorded in Punjab province 1171.5 mnds/hectare, the growth of yield in both province from 1998-99 to 2014-14, in Sindh province decrease 1% but the growth rate of yield in Punjab increasing 23% in same time period, in 2013-14 the yield of sugarcane in Sindh was recorded 1542.75 mnds/hac and in Punjab was recorded, 1443.75 mnds/hac, it means from 1998-99 to 2013-14 the growers of Sindh province stable same yield no big difference and Punjab province growers keep to

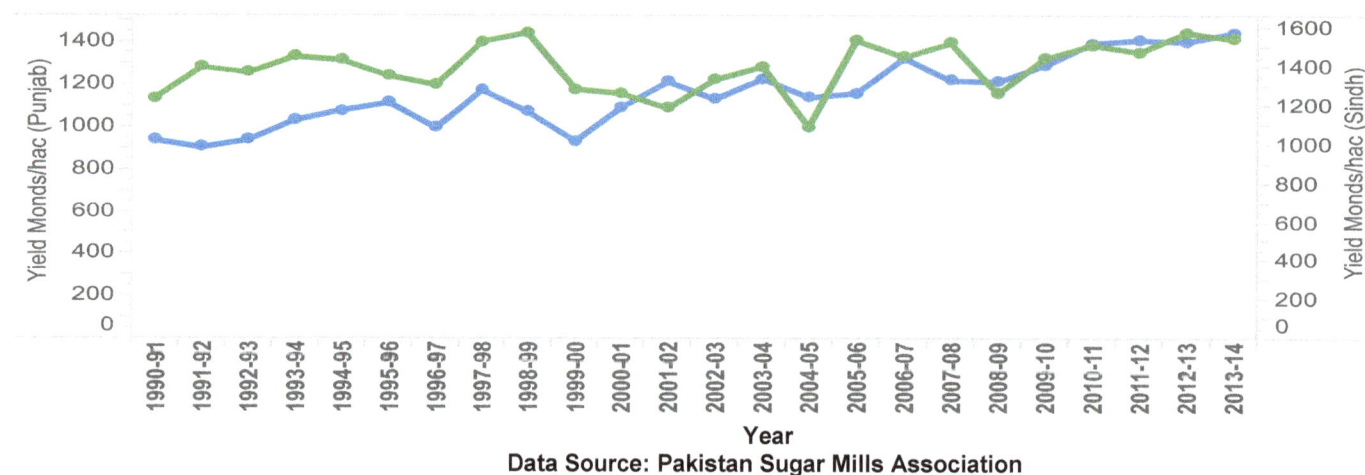

Data Source: Pakistan Sugar Mills Association

Figure 3. Sugarcane Yield hectare^{-1} with time index both Sindh and Punjab provinces of Pakistan

Table 1. Growth rate of sugarcane production, area and yield in Sindh and Punjab province of Pakistan, since 1990-91 to 2013-14

Year	Area (%)	Production (%)	Yield (%)
Punjab			
1990-91 to 1994-95	3.13	7.04	3.61
1995-96 to 2000-01	1.92	4.63	2.31
2001-02 to 2006-07	5.44	4.84	0.48
2007-08 to 2013-14	-1.52	2.18	2.92
Average growth 24 years	**2.24**	**4.67**	**2.33**
Sindh			
1990-91 to 1994-95	0.21	2.31	2.00
1995-96 to 2000-01	-0.66	-1.78	-1.58
2001-02 to 2006-07	6.13	12.07	5.96
2007-08 to 2013-14	0.01	0.78	0.73
Average growth 24 years	**1.42**	**3.35**	**1.78**

Data Source: Author's calculation with use of Excel (2016)

Table 2. Summary statistics of relationship between area and support price with production in Sindh and Punjab province of Pakistan

Independent	Coefficient	Standard error	t-test	Significant level	R^2	F
Punjab						
Intercept	1.189	0.335	3.553	0.002	0.97	175.68
Ln X_1	1.074	0.127	8.491	0.000	-	-
Ln X_2	0.159	0.022	7.212	0.000	-	-
Sindh						
Intercept	1.223	0.423	2.890	0.009	0.70	24.452
Ln X_1	1.184	0.175	6.750	0.057	-	-
Ln X_2	0.050	0.025	2.013	0.000	-	-

Data Source: Author's calculation with use of SPSS.

rising the yield, the variation of production and area, shortage of irrigation, climate changes, heavy flood, heavy rain, due to rain and flood insecticide and pesticide attack increased and the main factor of fluctuation production and area is marketing price of sugarcane Both Sindh and Punjab provinces of Pakistan.

Further analyze to growth rate of area, production with time index of sugarcane crop in Sindh and Punjab province of Pakistan. However, we tested average growth rate model for production, area and yield of sugarcane through the use of secondary time series data of sugarcane crop.

Table1 indicate that the growth rate of sugarcane production, area and yield in Sindh and Punjab province of Pakistan, results shows that in Punjab from 1990-91 to 1994-95 area, production and yield growth was increase 3.13%, 7.04% and 3.61 percent it was more than Sindh province was calculated 0.21% 2.31% and 2.00% respectively. In case of 1995-96 to 2000-01 was intended again more than Sindh province, in Punjab growth rate was, 1.92%, 4.63% and 2.31% but in this time period Sindh province area not increase but its goes minus -0.66%, production and yield also decrease -1.78% and -1.58%, it means in Sindh province have significant relationship between area, production and yield, if area fallen production and yield also fallen. Nevertheless in 2001-01 to 2006-07 the growth of area, production and yield of Sindh province increased much more than Punjab which was calculated 6.13%, 12.07% and 5.96%, in the same time period Punjab growth rate increased 5.44%, 4.84% and 0.84% respectively. In case of 2007-08 to 2013-14 in Sindh province growth rate of area, production and yield increased little 0.01%, 0.78% and 0.73% as compare to Sindh the growth rate of area in Punjab province decrease till -1.52% nevertheless the production and yield increased 2.18% and 2.92% it means in Punjab province have no big significant relationship between area yield and production. The overall average growth rate of sugarcane production, area and yield during 24Year since1990-91 to 2013-14 in Punjab growth rate increased more 2.24% of area, 4.67% production and yield 2.33%, as compare to Punjab province in Sindh province growth rate increased lower which was calculated 1.42% for area, 3.35 for production and 1.78% for Yield respectively.

Furthermore, the results of regression analysis which are presents in table2. As the cobb-Douglas production function was used, however the estimated co-efficient of the elasticity of the production, in case of Punjab the co-efficient of intercept is 1.189 representing the natural log of expected production of sugarcane when there is no impact of support price, the co-efficient of area (ln X_1) is 1.074 if the area of sugarcane increase 1 percent the production of sugarcane will increase 1.074 percent, and the co-efficient of support price (Ln X_2) also have significant which is 0.159, it measures if the support price increase one percent the production of sugarcane will increase 0.159%, in Punjab province both area and support price have significant relationship with production, the significant level is 0.1% percent, the value of R-square 0.97 which indicate that it's about 97% of total change in production of sugarcane crop, and the value of F-calculated is 175.68 which indicate that its highly

significant relationship between area and support price with production of sugarcane crop in Punjab, as compare to Punjab province the impact of supporting price and area on sugarcane production in Sindh province indicate that the coefficient of the elasticity of the production which is 1.223 indicate that the natural log of expected production of sugarcane when there is no significant impact of area and support price, after that the coefficient of area is 1.184 which indicate closely relationship with production, if the area increase 1 percent than the production will increased 1.184%, and the co-efficient of support price also have significant relationship but not highly level which is 0.050 it means if the support price increased 1percent thus the production of sugarcane will increase 0.050% in Sindh province, the significant level is 0.5% therefore the value of R-Square is 0.70 it means the possibility 70% total change of sugarcane production and the value of F-calculate is 24.452 it shows relationship of area and support price with sugarcane production in Sindh, but it is the lower than Punjab. due to water shortages, a lack of high yielding varieties and Mills are typically bound to pay, so the farmers do not willing to cultivate more area and use of more inputs. Therefore, it is necessity to support price increase if the support price increase than the farmers take keen interest for cultivating more area under sugarcane with use of modern technologies and also increase the applications of inputs, however the government of Pakistan should increase support price for promoting the sugarcane production both Sindh and Punjab province of Pakistan (Alam 2007)

CONCLUSION AND RECOMMENDATIONS

This study was conduct on the basis of secondary time series data since 1990-91 to 2013-14 for comparing the effect of support price to growth rate of sugarcane production in Sindh and Punjab province of Pakistan through the use of growth rate model and cobb-Douglas model. The results indicate that in Punjab growth rate of area, production and yield increased more 2.24% of area, 4.67% production and yield 2.33%, as compare to Punjab province in Sindh province growth rate increased lower which was calculated 1.42% for area, 3.35 for production and 1.78% for Yield respectively. However the relationship of area and support price with production have significant results shows, in case of Punjab the co-efficient of intercept is 1.189 representing the natural log of expected production of sugarcane when there is no impact of support price, the co-efficient of area (ln X_1) is 1.074 if the area of sugarcane increase 1 percent the production of sugarcane will increase 1.074 percent, and the co-efficient of support price (Ln X_2) also have significant relationship which is 0.159, it measures if the support price increase one percent the production of sugarcane will increase 0.159%, in Punjab province both area and support price have significant relationship with production, the significant level is 0.1% percent, the value of R-square 0.97 which indicate that it's about 97% of total change in production of sugarcane crop, and the value of F-calculated is 175.68 which indicate that its highly significant relationship between area and support price with production of sugarcane crop in Punjab, as compare to Punjab province the impact of supporting price and area on sugarcane production in Sindh province indicate that the coefficient of the elasticity of the production which is 1.223 indicate that the natural log of expected production of sugarcane when there is no significant impact of area and support price, after that the coefficient of area is 1.184 which indicate closely relationship with production, if the area increase 1 percent than the production will increased 1.184%, and the co-efficient of support price also have significant relationship but not highly level which is 0.050 it means if the support price increased 1percent thus the production of sugarcane will increase 0.050% in Sindh province, the significant level is 0.5% therefore the value of R-Square is 0.70 it means the possibility 70% total change of sugarcane production and the value of F-calculate is 24.452 it shows relationship of area and support price with sugarcane production in Sindh, yet it is the lower than Punjab. Both province results indicate that the production have significant relationship with area and support price its necessity to support price increase if the support price increase than the farmers take keen for cultivating more area under sugarcane with use of modern technologies and also increase the applications of inputs, however the government of Pakistan should increase support price for promoting the sugarcane production both Sindh and Punjab province of Pakistan. So that some policies recommendations for the improving of sugarcane production, in future it is important the research should be added as resource conversation with friendly environment objectives to ongoing agricultural research projects. New research should be provided to assist the farmers therefore they must be able to make better decision for practices management, the information about management should be provided through mass, electronic and paper media. Government should ensure the availability of new technologies with technical assistant, maintain canal irrigation system and also provide technically assistant to the controlling of pesticide and insecticide for rising production and supporting price also needed in both province Sindh and Punjab of Pakistan.

ACKNOWLEDGEMENT

The author acknowledged the contributions of Prof. Sergio Louro Borges for donating his time, critical evaluation, constructive comments, and invaluable assistance toward the improvement of this very manuscript.

REFERENCE

Adnan N, Jariko GA, Junejo MA (2013). Factor effecting sugarcane production in Pakistan. Pak J. comer. Soc. Sci. vol (1): 128-140

Alam SM (2007). Sugarcane production and sugar crisis. Economic review. http://findarticles.com/p/articles/mi_hb092/is_11_38/.

Azam M, Khan M (2010). Significance of the sugarcane crop with special and faience to NWFP. Sarhad J. agric., 26: 289-295

Deepchand K (1986). Economic of electricity production from sugarcane tops and leaves preliminary study. Ind. Sugar J. 88: 210-216

GoP Economic Survey of Pakistan (2015). Agricultural statistics of Pakistan. Ministry of Food agriculture and livestock division, Islamabad

LCCI (2013). An overview of sugarcane industries in Pakistan, publish by: Research development department, Lahore Chamber of Commerce and Industry.

Magsi H (2012). Support price: growth rate of cotton production in Pakistan. Agricultural journal 7 (1): 21-25

Masood, MA, Javed MA (2004).Forecast model for sugarcane in Pakistan, Pak J. Agri. Sci., Vol. 41(1-2).

10

Determinants of Contractual Choice and Relationship Sustainability in Organic Fruits and Vegetable Supply Chains: Empirical Evidence from Stakeholder's Survey from South India

*Ravi Nandi[1], Nithya Vishwanath Gowdru[2], Wolfgang Bokelmann[3]

[1]Program Manager, National Institute of Agricultural Extension Management (MANAGE), Hyderabad, 500030 Rajendranagar, Telengana State, India.
[2]Assistant Professor, National Institute of Rural Development and Panchayat Raj (NIRD&PR), Hyderabad, 500030 Rajendranagar, Telengana State, India.
[3]Department of Agricultural and Horticultural Economics, Invalidenstasse 42, The Humboldt University of Berlin, Germany.

Supply chain stakeholders for local organic food face uncertainties. In the present study, the empirical relevance of relationship types, farmers contracting choice and several determining factors which potentially influence both choices of contract types and the relationship with sustainability was tested. The study draws the Williamson's governance contractual structure of formal and non-formal relationship prevailing between chain actors to see what kind of contractual relationship is prevalent in the chain drawing on transaction cost theory. Data were captured by conducting a survey of 155 respondents (127 farmers, 11 processors and 17 retailers) belonging to Karnataka state in India. The analysis was both quantitative and qualitative and used binary choice models and Structural Equation Modelling (SEM) to analyse the key determinants. The analysis revealed that informal relationships were prevalent in the market. Market, sector and enterprise specific characteristics were found to influence the choice of the contract while dyadic, firm level factors influence relationship sustainability in the organic fruits and vegetable supply chains. Results have implications for agribusiness management and policy makers in relation to organic agribusiness development in the study area.

Keywords: Organic Fruits & Vegetables Supply chains, Relationship Sustainability, Binary Regression, Structural Equation Model, India

INTRODUCTION

The dynamics of agri-food supply chains and the globalisation process which has spread fast in recent years have resulted in dramatic changes in supply chains of developing countries, including India. The demand for environment-friendly food products such as organic food has significantly increased due to increasing awareness of health, food safety, and environmental concerns (Briz & Ward 2009). Consumer quality perceptions are based on individual evaluative judgments (Bredahl 2004). Quality traits are highly focused upon in the marketing of organic food products (Zanoli 2016). Consumers expect the food to be of high standard in terms of physical attributes, while process characteristics such as animal welfare, fair trade, and environmental ethics are also gaining importance across the globe. Furthermore, confidence in food safety is an important condition for consumers' buying decisions. Consumers' perceptions and concerns on food safety, quality, and nutrition are becoming very important across the world. One of the most accredited explanations assigns the main responsibility of the emerging interest in

*Corresponding author: Dr. Ravi Nandi, Program Manager, National Institute of Agricultural Extension Management (MANAGE), Hyderabad, 500030 Rajendranagar, Telengana State, India. E-mail: nandi999hu@gmail.com.

quality and safety issues to the various food scandals, and the consequential food scares that have emerged throughout Europe (Naspetti & Zanoli 2009). This has provided growing opportunities for the market of organic products in recent years. The untapped potential market for organic food in a developing country like India needs to be realized through organized marketing (Nandi et al. 2016). Marketers should be and obviously become became careful when claiming health benefits in order to motivate consumers to buy organic food because of the lack of evidence for this assumption (Lairon & Huber 2014). Despite this fact, beliefs related to health benefits are still revealed among consumers (e.g. Talamini & Révillion 2016). Addressing these consumers' needs calls for coordinated actions by companies throughout the chain (Ménard & Valceschini 2005). Therefore, the market success of products depends not only on each distinct stage of the supply chains but seem to be increasingly determined by collective strategies involving businesses throughout the whole chain. Thus, enhanced coordination among primary producers, processors, and retailers gains importance and the quality of their relationships is identified as a potential source of competitiveness, but more importantly, they expect the long-term relationships would increase their competitiveness over the years (Schiemann 2007). This is as a result of well-developed vertical coordination and can help to reduce business uncertainty, including that relating to food safety obligations, improve access to essential resources and result in higher business productivity (Dyer & Singh 1998). Therefore, there is an increasing acknowledgement that competition in the agri-food sector no longer takes place between individual companies but between entire chains or networks. The establishment and maintenance of relationships between partners are crucial, and it is increasingly important that partners build stronger and longer-term relationships in the supply chains to remain competitive because of the ever changing competitive environment (Parsons 2002).

Considerable research efforts have already been undertaken in the agriculture sector around the world to gain a better understanding of business relationships and thereby enabling their more effective management (Naspetti et al. 2011). For instance, (Schulze et al. 2006) investigated business relationships in the German pork sector and revealed a relatively low level of vertical coordination within this chain. Furthermore, (Alboiu 2012) investigated relation types and determinant factors in Romanian vegetable supply chains and revealed that informal contractual relationships are prevalent in the market, and the contractual enforcement is at stake. The study also highlighted the crucial roles of communication and personal relationships in the chain. Power imbalance and information asymmetry posed obstacles for the development of trust between chain actors in the Netherlands (Lindgreen et al. 2004).
In India, about 60 per cent of food quality is lost in the supply chains from farm to the final consumers.

Consumers end up paying approximately 35 % more than what they could be paying if the supply chains were improved, because of wastage due to improper handling as well as multiple margins and higher transaction costs in the current supply structure. The farmer in India gets around 30 % of what the consumer pays at the retail store. Compare this with the situation in the USA, where farmers can receive up to 70 % of the final retail price, and wastage levels are as low as 4-6 % (GOI 2013). This clearly demands supply chains research and from emulating those practices and tapping that expertise for supply chains in India. Although several studies were carried out in order to study the various agri products supply chains in India (Deliya et al.2013; FAO 2013; Reddy et al. 2010; Sharma et al. 2013), there are no empirical evidence studying the contractual relationships in organic fruits, and vegetables supply chains in India. India is the second largest producer of fruits and vegetables in the world. The Indian fruits and vegetable farm structure is characterised by a very large number of small holdings (about 80%) and a small number of large-scale farms.

The Indian organic food sector is in the early phase of its development. Presently, there are 0.55 million organic producers cultivating 1.10 million hectares in India (Willer et al. 2013). The area under organic farming is relatively small as compared to the total cultivable area in the country. Currently, fresh produce (fruits and vegetables) are the highest demanded organic food categories in the country. The organic fresh produce market is highly fragmented, and it is only concentrated in the big cities of India. As stated by the CEO of a large retail company 'there are problems with the supply chains of organic products in India, since all products are not always available.

Moreover, the price premium, which can be 50-70 per cent of regular prices, is certainly deterring consumers' (Mukherjee 2013). Designing and managing local organic food supply chains is complex, and it faces socially bound uncertainties such as poor collaboration, communication and information sharing (Kottila et al. 2005; Stolze et al. 2007). Improving organic supply chains is becoming of increasing interest to developing country like India, since the demand for organic products is growing steadily, providing market opportunities and premium prices for producers who comply with organic certification standards (Santacoloma 2012). The newly emergent organic produce supply chains around the globe have also been found to be excluding small producers due to reasons of high certification costs, smaller volumes they produce, and tighter control by chain leaders in the absence of local market outlets for the organic producers (Raynolds 2004). With this, the background aim of our study was:
1. To study the existing relationships types and contractual issues in organic fruits and vegetables (F&V) supply chains in Karnataka state, India.

2. To identify the key determinants of smallholder farmer contracting choice
3. To identify determinants of sustainable relationships between partners in organic F&V supply chains.

These objectives seem to be very relevant based on the existing literature, industry reports and experts opinions, as far as we know, there is no empirical work that has been undertaken with these specific objectives in the case of organic F&V supply chains' organisation in the Indian context and specific to Karnataka state which is one of the leading states in the country for organic food production.

This paper tested a set of hypotheses, and the first hypotheses deal with those factors that influencing the choice of relationship (contract) types in organic F&V supply chains and the second ware about those factors that lead to sustainable relationships among chain actors in organic F&V supply chains. To achieve the latter objective, a model is developed in which the main components that define the relationship of sustainability and communication quality and the important factors explaining the sustainability of relationships are defined. The paper is organised as follows: Section II presents the theoretical background and hypotheses and includes contractual governance structure and transaction cost theory. Section III explains the background of the study area and the organic fruits and vegetable Sector in India. Further, section IV presents the methodology, and Section reveals empirical findings & discussions followed by conclusions.

CONCEPTUAL ARGUMENTS AND THE EMPIRICAL EVIDENCE

Contractual structure governance and Transaction cost theory

According to Humphrey & Schmitz (2004), the main components of contractual governance structure refer to 'what to produce, how to produce, how much to produce and when to produce'. Hence, the governance structure refers to the relationship between firms/companies and the institutional tool using governance mechanism the explicit coordination is made, and the activities in the chain are performed. Vertically integrated markets may offer farmers the opportunity to produce and sell differentiated products with high value added (Gyau & Spiller 2008). However, the high standards of vertically coordinated markets impose challenges and barriers to farmers who do not meet the production and marketing systems' standards. Transaction costs (TC) are always associated with the exchange process, and the size of exchange determines the organisation form of the economic activity. Also, transaction costs are determined by the information asymmetry which may lead to the limited rationality and/or

opportunistic behaviour of one of the parties in the chain. Contractual relationships may offer some relaxation principles for these problems (Nadvi & Waltring 2004). As a result of positive transaction costs and limited rationality, the contract is suggested as an analytical frame (Hobbs 1996; Hobbs 2004; Williamson 2000). Hobbs (1996 & 2004) argued that in the case of traditional retail chains, cooperation and information exchange may contribute significantly to transaction cost reduction. Williamson (2000) reflected that the way a transaction is organised (e.g., spot or coordinated market) depends on 'rational economic reasons'. He suggested three main dimensions of these reasons: a) asset specificity; b) uncertainty; and c) frequency. Asset specificity refers to the degree to which a particular asset can have alternative uses; uncertainty is usually given by the incompleteness of contracts and given imperfect information, which can lead to the opportunism of one of the parties to an agreement; and frequency refers to the rate of repetition of a transaction.

Choice of contract types by smallholder farmers

Economic research indicates that, under certain circumstances, both suppliers and buyers can benefit from using contracts. The production, handling, and marketing of organic products follow different processes, as specified by the respective country standards. For products to sell as 'organic', their production and handling processes must be certified by an authorised certification body. Production and marketing activities between firms often require investment decisions by the parties involved. The policy makers are concerned about the protection of expected rents needed to compensate for incurred costs. Conditions in the organic sector suggest that chain actors may potentially benefit from contracts, given the growth in demand for organic products, inconsistent and short supplies, and the need for certification. In the present study, the four relation types mentioned above were grouped into two major contract types as a) Formal relationship (explicit) and b) informal relationship type (implicit).

Factors affecting relationship (contract) choice

Except for direct sales, farm products generally move along the supply chains with the assistance of handlers in the study region. Information asymmetry between stakeholders in organic food markets may exist due to highly differentiated supply chains, as a result of increasing competition and heterogeneity of consumers' demand. With respect to the product and process qualities of food products uncertainty may exist, as many quality characteristics are credence attributes that cannot be measured, even after consumption(Young & Hobbs 2002). The relation between contract type choice and product quality in agri-food chains has been provided in several empirical studies (Boger 2001; Raynaud et al.2005). The conceptual arguments and the empirical evidence lead to the inclusion of the following hypotheses

H1. Higher product quality and food safety influence the choice of formal (explicit) contracts.

H2. Quantity and delivery frequency influences the choice of formal (explicit) contracts.

H3. The price premium and payment mechanism influence the choice of formal (explicit) contracts.

H4. History and trust between the buyer/seller influence the choice of formal (explicit) contracts.

H5. The contractual penalties influence the choice of formal (explicit) contracts.

H6. The long-term oriented investment decisions influence the choice of formal contracts

Sustainability of contractual relationships

Recent empirical studies across the world suggested that sustainable relationships encompass qualities such as mutual trust, satisfaction and commitment (Lages et al. 2005). In addition, dynamic aspects (the evolution of repeated interactions and transactions over time) and consider non-coercive as well as coercive behaviour and past chain experiences (Lai et al. 2005). According to 'theory of repeated games'(Kandori 2008), one of the explanations of mutual trust is the interest of transacting parties in preserving the value of a reputation of honouring past promises. The value of such reputation increases with the time horizon of a relationship, and with the number of repeated transactions taking place between parties. Thus, more sustainable relationships require repeated transactions with a high frequency of interaction being essential for their success.

Determinants of Relationship Sustainability

There are different factors which potentially influence the relationship sustainability in the chain. Fischer et al. (2010) and Fritz & Fischer (2007) suggested that external factors as well as chain-internal, dyadic factors may be of relevance. The socio-economic and regulatory environment in which agribusinesses are embedded also exerts a significant influence on chain relationships (Hughes 1996). The existing literature on relationships in agri-food chains have cleared revealed the importance of communication, defined by the two dimensions, quality and frequency is an important influencing factor. Communication can be defined as the 'glue' that holds a relationship together (Mohr et al. 1996). Fischer et al. (2010) and Schulze et al. (2006) showed that access to up and downstream information enables retailers and suppliers to adapt to supply problems, and market changes more rapidly, thereby communication positively influencing relationship sustainability. From a TCE perspective, information sharing counteracts opportunistic behaviour and reduces adverse selection as well as moral hazard (Simatupang & Sridharan 2007). Taking into

consideration of the above findings the seventh hypothesis is defined as follows:

H7. Higher communication quality along the chain positively influences the relationship sustainability.

Several social phenomena have increasingly been acknowledged as factors affecting economic success. In agri-food chains, both business (prices, costs, and markets) and inter-personal (personal bonds, trust and friendship) aspects of relations were seen as being vital for chain performance (Hinrichs 2000; Winter 2003). Fischer et al. (2010) also revealed that factors affecting relationship sustainability are the existence of personal bonds and equal power distribution between buyers and suppliers. Furthermore, (Rodríguez & Wilson 2002) found how personal or social bonds influence relationship-building. They defined personal bonds characterised by familiarity, friendships and personal confidence which are incorporated in the relationship. These inter-personal ties are a form of social capital that enhances the maintenance of relationships. Therefore, the following hypothesis is defined to cover the relevance of social structures in a more general sense:

H8. Stronger personal bonds among chain partners positively influence the relationship sustainability.

Key people within an organisation are individuals who possess specific knowledge about their own business and the relationships with one or several important business partners. As such, it could be considered that the development of personal bonds by key people reflects a willingness to invest in a specific asset (human asset specificity) and hence signals a desirable trading partner. Therefore, concerning social structures, they play an important role. Key people leaving the firm to create a problem for many small and medium scale food enterprises and farms where appropriate succession arrangements have not been made (Fischer et al. 2010). In such a situation, successful continuation of commercial relationships with the partners may be a hindrance. Thus, we considered the following hypothesis to unveil the relevance of personal bonds in relationship sustainability:

H9. Key people leaving the firm negatively affect relationship sustainability.

Equal power distribution among chain partners increases the probability that commercial rewards will be distributed fairly among the partners. Boger (2001) revealed that, within the contractual arrangements of the Polish pig chain, farmers prefer to conduct business with buyers who cannot exercise bargaining power. In addition, Fischer et al. (2010) and Gracia et al. (2010) showed the importance of equal power distribution in the chain and its influence on relationship sustainability. Then the next hypothesis was defined as:

H10. An equal distribution of power among chain partners positively influences the relationship sustainability.

Organic products have been frequently associated with attributes such as traceability, local origin, and supply, small-scale units of production. In addition to prescribed standards for organic foods from organic certification standards, there are additional attributes like locally, and small-scale production is important attributes of organic foods. The degree of embeddedness of business in its local environment may support the development of sustainable relationships between this business and other local businesses (Lähdesmäki *et al.* 2009). Thus, to sustain organic food supply chains for a longer period of time, it has to consider the local environment. Thus, the hypothesis was defined as:

H11. *The higher the degree of embeddedness in the local environment in which chain and its partners operate, the higher the sustainability of relationships*

Background of the Study Area and Organic Fruits and Vegetables Sector in India

India produced around 1.34 million metric tons of certified organic products and exported 135 products during 2012-13 with total volume of 165262 MT. The organic products export realisation was around US $ 374 million registering a 4.38% growth over the previous year. Organic products were exported to EU, US, Switzerland, Canada, South East Asian countries and South Africa (APEDA 2014). The Indian domestic market for organic food has developed rapidly over the past few years. New processing firms, suppliers and organic stores are coming up on a monthly basis. Branded organic products have now made an entry into many conventional retail stores. However, most supermarkets and even many organic stores still do not sell fresh organic fruits and vegetables, either because of the logistical hurdles and high risks of selling perishable products or because no supplies are available (Osswald & Menon 2013). Farmers are practising sustainable small-scale agriculture lack adequate market access that allows them to sell their products profitably. At the same time, demand for fresh organic produce is highest among all product groups (Osswald 2010; Rao 2006).

Indian agriculture is the home of small and marginal farmers (80%), and there are about 121 million agricultural holdings in India, out of which 99 million were small and marginal. Therefore, the future of organic agriculture growth and food security in India is connected to the fate of this category of actors (Dev 2012). The smallholder farmers contribute around 70% to the total production of vegetables, 55% to fruits against their share of 44% in land area in the country (Dev 2012). Presently, there are 0.55 million organic producers cultivating 1.10 million hectares in India (Willer *et al.* 2013). The area under organic farming is relatively small as compared to the total cultivable area in the country.

Our study area, which is Karnataka state, is India's eighth largest state in a geographical area covering 1.92 lakh[1] km² and accounting for 6.3% of the geographical area of the country. The state ranks fifth in India in terms of total area under horticulture. It stands fifth in production of vegetable crops and third in fruit crop production. It is also the largest producer of spices, aromatic and medicinal crops and tropical fruits. Karnataka is also the second largest producer of grapes in the country and accounts for the production of 12% of total fruits, 8% of total vegetables and 70% of the coffee in the country. It is the third largest producer of sugar and ranks fourth in sugarcane production. Karnataka is highly progressive with regard to vegetable production and enjoys this advantage because of favourable climatic conditions without any extremes in temperature. Agriculture remains the primary activity and main source of livelihood for the rural population in the state (Bende 2013). Karnataka state the first state in the country to implement an organic farming policy. Realising the benefits of organic farming in contrast to conventional cultivation, farmers in Karnataka are increasingly becoming part of the organic movement. Benefits like a reduction in the use of external inputs, improvement in soil fertility, lower soil degradation, biological pest control and above all protecting the environment have become the driving force of this movement. Following is the information on the status of organic farming and organic stakeholders in Karnataka State as per the available data collected by Research Institute on Organic Farming, Bangalore during 2014 (Devakumar, 2014).

Table 1: Status of Organic farming in Karnataka state, India

Total area under organic certification (2010 - 11 data)	80,706 ha
Total number of certified organic farmers	16,432
Number of operators/processors and exporters of organic produce/products	47
Number of Private Organic Outlets/Retailers	83
Number of organic Restaurants in Bangalore	10
Marketers of Organic Produce/Products	19
Number of NGOs involved in the promotion of organic farming in the state	129
Number of organic Farming Research Institutes	08
Number of certified Organic Operators	246

The market for organic produce only concentrated in major cities of India. In Karnataka, Bangalore is a major market for organic products. Fresh produce is transported to the

---

[1] 1 Lakh = 100000

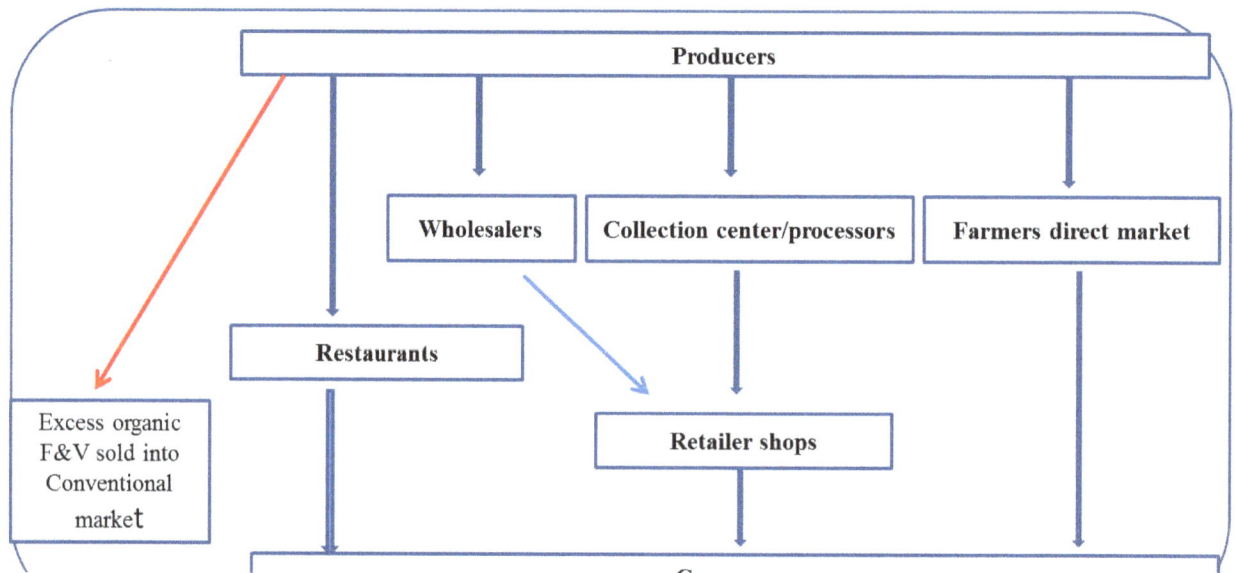

Figure 1: Simplified organic fruits and vegetable supply chain in Karnataka, India.

city from the peri-urban areas of Bangalore. The city was purposively selected for the study based on the report by Oswald (2012) in collaboration with International Competence Centre for Organic Agriculture (ICCOA) in 2012. This report revealed that 'among the three urban organic markets analysed, Bangalore is the largest urban organic market in India, and there is no sufficient information about organic stakeholders and their role in supplying products to megacities of India.'

The organisation of the organic supply chains does not differ much from the conventional chains; however, to maintain the organic quality attribute, stronger vertical coordination with sustainable relationships among the actors, as well as clearly defined actors roles and responsibilities are very much needed. Organic F&V supply chains in study area range from very short ones where farmers market directly to local consumers, to more complex chains where a number of different actors are involved in bringing organic products from farm to fork while keeping the organic quality attributes. The network (map) of the domestic organic fruits and vegetable supply chains in the study area are as shown in Figure 1.

METHODOLOGY

To test the hypotheses, the relationship situation between the chain actors in the organic fruits and vegetable supply chains in the study area were analysed. Here, fruits referred to only selected ones (Banana, Mango, Sapota, Grapes, Guava Papaya, Watermelon, and Jackfruit) which are grown by the certified smallholder organic farmers in the study area. The reason for selecting above mentioned fruits was that these are the most common organic fruits grown and available in the market and as organic food market is in the early phase of its development in the

region, there is no statistics regarding the share of individual fruits in total production of fruits production in the State. The relationship in the chain studied divided into two chain stages. The first chain stage represents the upstream between smallholder producers and processors; the second stage is the downstream relationship between processors and retailers. Processors here refer to fresh produce handlers who procure fresh produce from the smallholders and grade them on the basis of quantity, quality, and size, packing, labelling and branding of produce. Branding is done in only selected fruits and vegetables.

Data collection

The paper is based on data provided by 127 smallholder certified organic farmers, 11 processors and 17 retailers/supermarkets located in and around Bangalore city, Karnataka state, India. The source of information used in this study was mainly obtained from personal interviews based on the structured interview schedules, carried out on the total of 155 sample respondents (127 farmers, 11 processors and 17 retailers). Purposive random sampling was drawn from an official list from the state department of agriculture during Nov. – Dec. 2013. The validity of the interview schedules was assessed by a panel of experts from the state department of agriculture, experts from NGO and industry experts in the state. Reliability of the scales of the questionnaire was also computed by Cronbach's Alpha method and the coefficients of Cronbach's Alpha, which are appropriate for the study (Gliem & Gliem 2003). Farmers selected for the survey were smallholders having less than 2 ha of agricultural land and cultivating fruits and vegetables. Regarding the interviews with representatives of supermarkets/retail chains, they were selected randomly based on their willingness to answer questions related to issues regarding the procurement method and the structure of contractual governance with reference to product quality & quantity, frequency of delivery, organic food standards, price and payment mechanism, premium for organics, contractual penalties were asked. Before random selection of representatives based on the willingness to answer the questions, supermarket/retailers are randomly selected from the city, and then representatives were approached from the randomly selected supermarket/retail chains. The analysis was both quantitative and qualitative and took into consideration stakeholders responses to the questions asked concerning the relationship type and contractual aspects along with the set of questions about producers contracting choices. Open comments were also introduced in the interview schedules. The questionnaire design and responses were analysed by employing the structure proposed by Williamson. Further, the binary logistic model was used in order to identify the determinants of organic farmers' contracting choice.

To assess the factors which potentially influence the contracting choice of smallholder organic farmers a five-point Likert scale was used, where 1=total disagreement and 5=total agreement. The statements (variables) included are the 'importance of the fulfilment of organic food standards', 'the importance of price premium and payment mechanism', 'the importance of frequency and quantity delivered', 'the importance of trust in the chain partners', 'the investment decision' and 'the role of contractual penalties'.

The statements selected were based on the previous literature (Alboiu 2012; Fischer *et al*. 2010; Gracia *et al*. 2010; Lähdesmäki *et al*. 2009) after having discussion with panel of experts formed by industry experts, researchers and technicians from the Agricultural University and independent research organization in the state, in order to decide on the relevance of questions to be included in the final questionnaire for the survey. We used as the dependent variable, the type of contract they used. The dependent binary variable uses the value one if the contract is formal and 0 otherwise.

In order to find out the probability of selecting a certain type of contract, the frequency and quantity delivered was used as an independent variable in the model. The quantity delivered has an important role in selecting a type of contract. Normally, large commercial farms, which produce in bulk, prefer to have written contracts, while small farmers prefer informal/oral contracts. Price and payment mechanism is next independent variable used in the model, and it is very important as farmers usually choose the type of contract based on the price premium for organic products and payment mechanism i.e. on the same day or some days after delivery. Further, food safety and organic quality standards represent important conditions mentioned in the formal/written contract, and these conditions are also taken into consideration in oral contracts. The contractual penalty is another important variable taken into consideration in the model. Normally we see this in written contracts. According to institutional and transaction cost theory, a contract is not considered complete due to limited rationality. Furthermore, the trust and investment decision by farmers are important variables considered in the model. Long-term oriented producers who plan investments are likely to select a formal contract, and it is an important step to creating a reliable relationship (Alboiu 2012). Generally, history and trust between the partners influence the more informal contracts.

Binary Logit Model

A binary logit model (Cramer 2003) was used to determine the factors influencing farmer's contractual choice. In the present study, the observation unit is the individual farmers.

Empirically the logistic model represented as:

$$log\left\{\frac{p_i}{1-p_i}\right\} = \beta_0 + \sum_{j=1}^{k} \beta_{ij}x_{ij}; \qquad (1)$$

where, $p_i = prob(y_i = 1)$ and the left hand side corresponds to the logit or the log of the odds ratio. In the present context, the noticed y_i^* dummy variable can be defined as desire or probability to choose formal contract, farmers contractual choice as a dependent variable on the left-hand side and it is measured by dichotomous variablea which takes the value one for formal contract and zero otherwise. The final model was presented as follows:

$$log\left\{\frac{p_i}{1-p_i}\right\} = \beta_0 + \beta_1(Quality\ \&\ safety) +$$
$$\beta_2(quantity\ \&\ delivery\ frequency) +$$
$$\beta_3(price\ premium\ \&\ payment\ mechanism) +$$
$$\beta_4(Trust\ \&\ history) + \beta_5(Penalty) +$$
$$\beta_6(investment\ decision) \qquad (2)$$

Structural Equation Modelling (SEM)

SEM in its most general form consists of a set of linear equations that simultaneously test two or more relationships among directly observable and/or unmeasured latent variables (Bollen 1998; Fischer *et al*. 2010). In technically it can be defined as

$$X = \Lambda.\xi + \delta, \qquad (3)$$

Where **x** is a vector of indicator variables, Λ is a matrix of factor loadings, ξ a vector of latent factors and δ a vector of measurement errors. Under suitable and fairly general assumptions, the Covariance matrix $\sum$ of the observed variables **x** can be expressed by the three parameter matrices, Λ, Φ and Θ_δ:

$$\sum = \Lambda\Phi\Lambda' + \Theta_\delta, \qquad (4)$$

Where Φ and Θ_δ are the covariance matrices of factors ξ and measurement errors δ, respectively.

The objective of SEM is to estimate the unknown elements of these matrices (i.e. the missing model parameters) such that the covariance matrix generated by the model:

$$\hat{\Sigma} = \sum(\hat{\Lambda}, \hat{\Phi}, \widehat{\Theta_\delta}) \qquad (5)$$

Reproduces the empirical covariance matrix as exactly as possible. We used the STATA 12 software package with unbiased covariance as the input matrix. Given the existence of missing values in the dataset, maximum likelihood estimation was applied.

As mentioned above, the statistical analysis of data consisted of two parts. The first part, factors affecting the smallholder farmer's contract choice was identified, using a binary logit model. Logit regression allowed the prediction of a discrete outcome, such as group membership, from a set of variables that may be continuous, discrete, dichotomous, or a mix of any of these (Cameron & Trivedi 2005). Second, factors which influence relationship sustainability were analysed using Structural Equation Model (SEM). SEM approach is used to empirically test the influence of communication quality and other four predictors on relationship sustainability in organic fruits and vegetable supply chains.

This approach is chosen because of the analysed concepts, relationship sustainability and communication quality cannot be directly observed, but can be considered unobserved (latent) variables measured by one or more indicators. However, SEM permits the analysis of simultaneous relationships between dependent and independent variables affecting relationship sustainability. Structural equation modelling consists of an entire family of models where the multiple and interrelated dependence relationships are estimated, and unobserved concepts are represented in these relationships (Hair *et al.* 2001). The data analysis procedure consists of a principle component analysis (PCA) to assess the measurement model and the SEM analysis to examine the overall relationships among the constructs (Hair *et al.* and Black 2001).

Two constructs (communication quality and relationship sustainability) were used in this analysis. All other variables were measured as single items (see Table 7 for a description of employed variables and items). Construct reliability (Anderson & Gerbing 1988; Fischer *et al.* 2010), as assessed by Cronbach's alpha (**0.8**1), was regarded as satisfactory. Construct validity was assessed using principal component analysis (PCA) on the four items. Only one principal component could be extracted (representing **61%** of total original variabilities), thus demonstrating the construct's uni-dimensionality. For the 'communication quality,' construct two items were used, with the Cronbach's alpha 0.86, this construct was also regarded as satisfactory. Knowing the only two items in this construct, PCA is not meaningful for this construct. Upon performing PCA on both the constructs (six items) resulted in two components extracted. The first constructs four relationship sustainability items, accounting for 39 per cent of total variability, and the second construct consisted of two communication quality items, representing 31% of the original variability.

RESULTS AND DISCUSSION

Contractual relationship types

Williamson (1991, 2000) categorised relationship types into two, namely formal and non-formal. The merits of formal relationship (written type) have been identified by many scholars (Bullington & Bullington 2005; Young & Hobbs 2002). Their findings revealed that they help to secure high product and process quality and to implement food safety standards and controls. Effective business relationships can help to reduce uncertainty along the chain by securing a more stable flow of orders, and it contributes to better access to important resources results in higher productivity (Dyer & Singh 1998). In the present study certified smallholder organic fruits and vegetable growers were asked to present what type of contractual relationships they used in their business. Table 2 shows the relationship types for the two chain stages. The result revealed that the percentage of the formal/ written relationship was extremely low at both the stages (farmer-processor, processor-retailer). Comparatively, a higher percentage of the formal relationship was observed at the downstream (Processor-retail level) of organic F&V supply chains. Supermarkets/retailers are likely to select more formal relationship type with processor or supplier, in comparison with the farmer. Showing that, downstream businesses are more likely to organise their relationships more systematically and in a standard way. The results were in line with studies in Romania, Europe and USA (Alboiu 2012; Dimitri 2010; Fischer *et al.*2008) but the difference is the per cent of formal relationships at farmer's level in the present study were much lesser than those mentioned. This may be due to the early phase of the organic market in India and organic supply chains from a smallholder's perspective not yet organised or linked to the market. The majority of the smallholders are being excluded from the supply chains due to the reasons of high certification cost, smaller volumes they produce, and tighter control by the chain leaders in the absence of adequate retail outlets for organic (Raynolds 2004).

Table 2: Relationships choice by smallholder organic fruits & vegetable farmers

Relationship Type	Supply chain Stakeholders				Average %
	Stage I		Stage II		
	Farmer -->	Processors <--	Processors -->	Retailers <--	
Formal	11 (8.66%)	1 (25%)	4 (57%)	6 (35%)	31. 40
Informal	116 (91.33)	3 (75%)	3 (43%)	11 (65%)	68.52
Total (*n*)	127	4	7	17	100 (*n*=155)

Note: *Based on the field survey 2012/13*

Factors influencing the choice of contract types

In order to analyse the determinants of the relationship type choice, the binary logit model was used, where the formal relationship was 1, and the informal relationship was 0. The farmers were asked to rank from 1 to 5 on a Likert scale, 1 being total disagreement and 5 being a total agreement to scale statements referring to quantity and frequency of delivering, price premium and the mechanism of payment, quality standards requirements, trust and history of relationship in the partners, contractual penalties and farmers investment decision. The model included six variables, each one associated with the hypotheses as discussed above. For a detailed description of the variables see Table 6. The results are as shown in Table 3. From the variables hypothesised to identify the determinants of choice of contractual type, five variables had statistically significant effects. Negative (positive) parameters estimated indicate that an increase in the value of the independent variables corresponds to decreasing the (increase) probability of choosing a formal/written contract type instead of an informal type. As per our expectations, organic fruits and vegetable quality and safety orientation (H1) seems to increase the likelihood to choose a formal/ written contract. One of the most important factors driving Indian urban consumers towards organics is food safety and assured quality for organic products (Ravi 2014). Thus, to get certified organic fruits and vegetables, supermarkets encourages the use of a written contract with producers. However, Raynaud et al. 2005 have revealed, depending on the quality assurance scheme (private brand or a public certification process such as a Protected Geographical Indication) the preferred governance structure might well be different (more spot markets in the case of PGIs and more formal/explicit contracts in the case of private brands). So, from this point of view, our results were in support of private brands. Further, quantity and delivery frequency have an important influence upon formal contract choice, and it was highly significant, thus confirming H2. Generally, larger farmers were more likely to choose written contracts than smaller ones, in the case of organic food production, due to niche market it always advantages for the producers to have assured demand from the supermarkets to produce in bulk and supplied it to buyer frequently. The estimated parameter coefficient for the price premium and payment mechanism was positive and significant, indicating that the higher price premium and more convenient payment mechanism, higher the probability of signing a written contract, hence confirming H3. It was found that at present situation, supermarkets/specialised retailers normally make payments for the farmers 7-15 days after the transaction takes place without providing any incentives for them. This situation makes difficult for the smallholders to manage their livelihood as most of the cases agriculture is the only source of income for them. Contrary to our expectations, history and trust between farmers and their partners play a significant role in selecting informal contracts. Longer history and higher trust among the partners lesser the probability of selecting formal contract, thus rejecting our hypotheses (H4). The variable contractual penalty was not statistically significant, but the sign of the estimated

coefficient is as we expected (H5). Finally, variable investment decision by the farmer was positive and significant, thus confirming the hypotheses H6. Generally, long-term orientation has vital step to creating a reliable formal contractual relationship for planning and securing future supply or sales. In other words, longer-term oriented organic farmers who plan investments are more likely to choose a formal contract. Our results support the findings of Schulze et al. 2006 for Germany and Fischer et al. 2010 for European agri-food chains, Gracia et al. 2010 for Spanish wheat to bread chain and Lindgreen et al. 2004 for the Netherlands. All these studies focused on conventional food chains as against organic F&V chain in our case. However, as per our knowledge, there are no studies with similar objectives in India for organic food supply chains, making our study relevant for the development of organic fresh produce market in the area.

Table 3: Binary logistic regression results: Estimated Factors influencing smallholder farmer's contractual choice

Variables	Parameters Estimated (β)	Hypothesis status
constant	-10.90***	
Organic quality and safety standards	1.05**	Accept H1
Quantity and delivery frequency	1.44***	Accept H2
Price premium and payment mechanism	1.08***	Accept H3
History and trust with partners	-1.4***	Reject H4
Contractual penalties	0.46	-
Investment decision	0.91***	Accept H6
Model statistics		
Percentage of formal contractual type correctly predicted	48.4	
Percentage of informal contractual type correctly predicted	91.7	
Overall percentage correctly predicted	81.10	
Nagelkerke R²	0.50	
Cox & Snell R²	0.33	

Notes: Formal contract type=1, Informal contract type=0, statistically significant at the 1 %(***), 5% (**) and; sample size: 127

Relationship Sustainability Index

Relationship sustainability index (RSI) scores are mentioned for the organic F&V supply chains for two chain stages. Respondents were asked to rate their opinions, on a scale from 1= very poor, to 5= very good. The following statements concerned with the quality and stability of their important supplier/buyer relationship were used;
*Our trust in our supplier/buyer. * Our commitment towards our buyer/supplier.
* Our satisfaction with our buyer/supplier and * our collaboration with our buyer/supplier in the past.

These scores related to the respondent's relationships with their important suppliers or buyers only. The RSI scores are as shown in Table 4. RSI was calculated as an unweighted average of the scores obtained for the above mentioned four statements. RSI scores were calculated only when valid data on each statement were available. As reported in Table 4, Relationship sustainability scores are higher in the downstream relationship (Processor-Retailer) with a mean score of 3.49 as compared to upstream (Farmer-Processor) which was 3.10. The differences between downstream and upstream relationships were statistically significant (at $p < 0.05$). Relationships sustainability varies between the chin stages with downstream relationships being generally perceived as more sustainable than upstream. These results were in line with Fischer et al. (2010) and Gracia & Albisu (2010), who analysed sustainable relationships in four pig meat and cereals supply chains in four European countries, and RSI scores differ mainly across chain stages and only to a limited extent between the different agri-food chains.

Table 4: Relationship sustainability Index scores in organic fruits and vegetable supply chain in Karnataka State, India.

Chain stages						
Organic F&V supply chain	Farmer-processor			Processor-Retailer		
	Mean	SD	n	Mean	SD	n
	3.10	0.80	138	3.49	0.62	28

Source: Based on the field survey 2012/13

Factors which influence relationship sustainability

The factors which potentially influence relationship sustainability in organic F&V supply chains were analysed by estimating the standard structural equation model (SEM), which was used to examine the general fit of the model proposed and to test the hypotheses formulated. The SEM estimated results for the data are presented in relation to the hypotheses concerning the relationship sustainability in organic F&V supply chains. The pooled SEM estimation results are shown in Figure 2. Furthermore, a numeric summary of the results from the structural equation model are presented in Table 5. The model fitted the data well, with all goodness-of-fit measures below (above) the recommended acceptance levels (CMIN/DF=1.78, RMSEA= 0.045, CFI=0.93 and TLI=0.98). Bollen (1998); Hair (2009) reported, CMIN/DF values should be between 1 to 3 (closer to 1 is better), root mean square errors of approximation (RMSE) must be less than 0.05 and TLI/CFI (Turker Lewis Index/ Comparative Fit Index) should be ≥ 0.90 60. Overall 61 % of the variance in the observed relationship sustainability construct can be explained by the determinants identified. In the structural model, five variables were found to have a positive and statistically significant impact on the relationship sustainability construct: communication

quality, personal bonds, effects of key people leaving the firm, equal power distribution between chain partners and local embeddedness. Thus, confirming our hypotheses H7, H8, H9, H10 and H11 explained in section 2.5. The path diagram for the estimated model is as shown in Figure 2. This figure represents the latent variables as oval and indicators as rectangles.

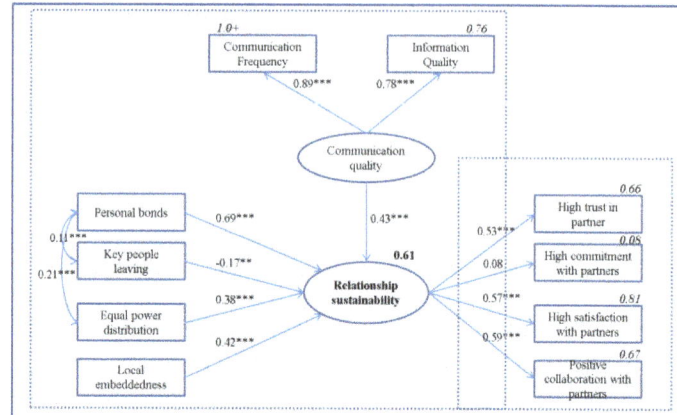

Figure 2: Path diagram of the estimated generic model: Determinants of relationship sustainability- pooled SEM estimation results.

Based on the SEM model presented in Table 5 and figure 2, the most important influencing factors to the Relationship sustainability construct is the existence of personal bonds between the chain partners followed by communication quality with standardised regression weights of 0.69 and 0.43 respectively. These results were in line with previous literature in both developed and developing countries. These finds that the existence of personal bonds is particularly important in relationships with farmers while effective communication is more crucial in relationships with retailers (Fischer et al. 2010). Furthermore, Gracia et al. (2010) described personal bonds do not influence directly the quality of relationships between chain partners, but through communication quality, it influences relationship sustainability. The personal bonds positively influence the quality of communication indicating that communication improves as personal bonds are closer along the chain. In addition, (Mohr et al. 1996) revealed, communication was one of the most important influencing factors in achieving successful inter-organization cooperation. Effective communication is an indispensable step for the creation of trust along supply chains and hence for the management of sustainable economic relations (Greenberg & Graham 2000; Lofstedt 2006). Chain local embeddedness is the third important factor (0.42) which influences relationship sustainability. Chain embeddedness is very important in organic supply chains as compared to the conventional one, as products produced in the local area by smallholders are attributes of organic food. Sustainable economic relationships is influenced by local embeddedness and it is constituted of four indicators: whether the products of the firm are Local

Table 5: SEM estimation results with standardized parameters[1] and significance levels.

		Estimates
Structural Model	Relationship sustainability <-- Communication quality	0.43***
	Relationship sustainability <-- Existence of personal bonds	0.69***
	Relationship sustainability <-- Equal power distribution	0.38***
	Relationship sustainability <-- Key people leaving organization	-0.17**
	Relationship sustainability <-- local embeddedness	0.42***
	Frequency <-- Communication quality	0.89***
	Quality of information <-- Communication quality	0.78***
	Key people leaving organization <--> Existence of personal bonds	0.11***
	Equal power distribution <--> Existence of personal bonds	0.21***
	R^2 Relationship sustainability	0.61
Measurement model	Frequency of communication <-- communication quality	1.0+
	Quality of information <-- communication quality	0.763***
	Trust <-- Relationship sustainability	0.534***
	Commitment <-- Relationship sustainability	0.077
	Satisfaction <-- Relationship sustainability	0.575***
	Collaboration history with partners <-- Relationship sustainability	0.598***
	R^2 Communication frequency	0.669
	R^2 Quality of information	0.766
	R^2 Trust on partners	0.666
	R^2 Collaboration history with partners	0.673
	R^2 Commitment	0.079
	R^2 Satisfaction	0.815
Overall fit	Normed X^2=CMIN/DF	1.78 (p=0.168)
	RSMEA	0.045
	CFI	0.03
	TLI	0.98

Notes: [1] In the structural model, <-- are regression weights and <--> are correlation coefficients; in the measurement model <-- are factor loadings, --> are regression weights and <--> are correlation coefficients. R^2 are squared multiple correlations in the structural model and communalities in the measurement models. *** (**) means statistically different from zero at the 1% (5%) significance level.

Products, whether the firm's suppliers were from the local area, whether their buyer was from the local area and whether the firm participates in the local community (Revoredo-Giha et al. 2010). Equal power distribution between the chain partners was the fourth factor (0.38). Power distribution plays a significant role in the nature of economic relationships; power distribution in economic relationships is determined by many factors, such as relative market share, information asymmetry (Lähdesmäki et al. 2009). Finally, the variable key people leaving the firm (-0.17) had a negative significant influence on relationship sustainability. The variables existence of personal bonds and key people leaving firm were positively and significantly correlated with each other, suggesting that key people are those who develop personal bonds with business partners. Further, the existence of personal bonds and equal power distribution are also positively and significantly correlated with each other. From a theoretical perspective, there is no reason to believe that when key people leave a company the power distribution between that company and its buyer or suppliers changes. There could be a deterioration of communication quality as a result of key people leaving the firm.

Based on the measurement model, relationship sustainability construct and communication quality performed well, with all factor loadings being greater than recommended level 0.50 (Hair et al. 1998) and communalities also being greater than 0.50 except for the variable commitment, which was 0.08. In the relationship sustainability construct, the most important components are a positive collaboration with partners, satisfaction, and trust. Collaboration can be direct or indirect. Firms may interact indirectly by sharing a common infrastructure (e.g., market), or collaborate directly through establishing contact with a potential partner and receiving a reaction. Here, the focus is on direct interactions. Collaboration history comprises all positive and negative experiences made with the exchange partner and is used as a basis for deciding on future actions with the exchange partner. When firms transact for the first time, they generally have no experience with an exchange partner and are limited in their evaluation possibilities (e.g., with regard to a partner's trustworthiness). This may not be important for arm's-length transactions, such as in spot markets, adverse selection can cause severe hold-up problems at critical phases in longer-term relationships. Firms which have a positive collaboration history are characterised by economically rewarding transactions, for all involved parties, successful, productive endeavours, and critical phases which have been endured and successfully resolved. Therefore, a positive collaboration history influence to the sustainability of relationships by reducing the probability of partners switching to other buyers or suppliers (Anderson & Weitz 1989; Bejou et al. 1996). Further, the ability to achieve high levels of relationship satisfaction has been considered an essential ingredient of business success(Morrissey & Pittaway 2006), because satisfaction will affect the morale and subsequent intentions of business partners to participate in joint activities (Schul etal. 1985). Thus, building a satisfactory relationship between buyer and seller is crucial for both farmers and buyers. Trust may be that quality assurance and safety procedures, along with an acknowledgement of competence in such matters, contribute to the development and importance of such trust (Lindgreen 2003). Finally, the frequency of communication and quality of information are equally important for the 'communication quality' construct.

CONCLUSIONS

In this paper, existing contractual relationship types, determining factors for relationship choice and factors influencing relationship sustainability in organic fruits and vegetables supply chains in the Karnataka State of India were analysed. The results were based on both quantitative (majorly) and qualitative analyses on 155 respondents along the organic F&V supply chains.

Knowing the fact that Indian organic food market is in an early phase of its development, our results about existing relationship (contract) type among the chain partners revealed that, there is a high degree of uncertainty between stakeholders both in terms of contractual relationships and contract enforcement along the supply chains. Number of formal contracts are lesser (31.40%) than informal ones (68.52%). In addition, the share of formal relationships is higher in downstream level (processor-retailer) as compared to the upstream (farmers-processors) level. This situation may lead to higher uncertainty at the producer level.

The relationship choice determining factors are majorly quantity & delivery frequency, as smallholder farmer considers their role only when they allow the security of their sales, specifically from the point of investment to produce higher quantity. Based on the price premium and payment mechanism hypotheses, they influence positively to formal relationship choice. When the price, quantity, frequency, and payment conditions are fixed in the contract, an increase in market price will increase the benefits for the producers as compared to selling in the conventional market (an open market which is outside of contract). Further, safety and quality are the main attributes of the organic products. When producers fail to deliver the prescribed quality of products due to the absence of a formal contract with fixed conditions, it will deteriorate the downstream level relationships. The chain partners having long-term relations and trust influences the informal relationship. Thus, the formal contract may

ensure consistent quality, supply, secure products in limited supply and more investment by the stakeholders this ultimately leads to the development of the sector under study.

Several factors have been identified as important determinants of relationship sustainability in organic fruits and vegetable supply chains, all of which can be managed at the organisation level. Regarding the relative importance of the relationship sustainability determinants, the SEM estimation found out chain local embeddedness as being crucial followed by communication quality. Communication quality was the most important in most of the studies mentioned here, but in our case, local embeddedness being most important. Thus, considering strong regional or local identity of produce and local activities outside the business operations positively affects the relationships sustainability.

The market power asymmetries between business partners, often due to differences in the scale of the firms, can create a feeling of insecurity and vulnerability among small partners in the chain. Therefore, power asymmetries can reduce trust and commitment and can be harmful to the relationship sustainability. Hence, it is acknowledged that where there is fair treatment, the effect of unequal power distribution may be reduced. In addition, the effect of key people leaving the firm is from a theoretical point of view related to the importance of personal bonds, and these two factors are positively and significantly correlated with each other. This indicates that key people generally are those who maintain business relations and develop personal bonds with the partners. The effect of key people leaving on relationship sustainability has been consistently estimated as negatively influenced, but it is not always significant and normally low in magnitude.

Our results combined with qualitative analyses suggest that small holders prefer formal contracts because they can deliver larger quantities. But the situation is not favourable as there is less chance of sales for the quantity they are producing leading to selling organically produced fruits and vegetables in the conventional market as this produce are perishable in nature and also price premium they get not satisfactory. An organic quality and safety standards were significantly influences the formal contract choice, and history and trust among the partners were negatively contributed to the selection of formal contract, the investment decision positively contributed to the selection of formal contract.

Overall, results can be concluded that there is an increased uncertainty along the chain and it is more uncertain in upstream level (farmers-processor) in terms of what kind of F&V needs to be produced, how much to be produced and where to sell. Thus it is negatively influencing smallholder revenue and their investment decision. Given the significant differences of relationship sustainability between chain stages, improvements must be targeted at the upstream level as compared to downstream level. This may be challenging due to farmer's dispersion over the larger geographical area and lower level use of information and communication technology adoption at farmers level. Further, enhancing technological and human communication capacities may lead to higher information quality and more transmission frequencies may help to build and maintain stronger fresh produce supply chains. From a managerial point of view, personal dimensions that potentially significant for its relationship sustainability. Thus, these dimensions need to be acknowledged by managers as it can help to improve relationship stability by strengthening mutual trust and commitment. From a policy perspective, our results showed that policies that focus on improving business to business communication are likely to have a positive influence on relationship sustainability.

Future research in this area may consider other organic food supply chains that have discrete production periods. In our analysis, we have focused on fruits and vegetables (few selected) which are continuously produced products. Also, large-scale longitudinal studies on the development of vertical and inter-firm relationships in organic food supply chains also are valuable, and it can help to understand how relationship sustainability develops over a period of time.

Table 6: Variables in the logit model

Variables	Description	Likert Scale (1-5)	n	Mean	SD
Contract type*: Formal/informal	Type of contract they used	1=Formal contract, 0 otherwise	127	0.24	0.43
Organic quality and safety standards	Quality and safety standards influence the choice of contract	Likert scale: total disagree..., total agree	127	2.30	061
Quantity and delivery frequency	Quantity and delivery frequency influence the choice of contract	Likert scale: total disagree..., total agree	127	3.19	0.85
Price premium and payment mechanism	Price premium and payment mechanism influence the choice of contract	Likert scale: total disagree..., total agree	127	2.35	0.75
History and trust with partners	History and trust with your partner influence the choice of contract	Likert scale: total disagree..., total agree	127	3.47	0.83
Contractual penalties	Contractual penalties influence the choice of contract	Likert scale: total disagree..., total agree	127	3.25	0.79
Investment decision	Investment decision influence the choice of contract	Likert scale: total disagree..., total agree	127	3.14	0.87
*Dependent variable is, the type of contract farmers used with the buyer. The dependent binary variable takes on the value 1 if the farmer had formal contract and 0 otherwise.					

Table 7: Items/indicators used in the SEM model

Variables	Description	Likert Scale (1-5)	n	Mean	SD
Personal bonds	The relationship between the partners is characterized by strong personal bonds	Likert scale 1 to 5, 1 being totally disagree and 5 being totally agree	155	2.88	0.81
Communication frequency	Satisfaction with the communication frequency of our supplier/buyer in our important business relationship	Rating scale: total disagree..., total agree	155	3.14	0.84
Information quality	Satisfaction with the quality of information received from our supplier/buyer	Rating scale: total disagree..., total agree	155	3.12	0.93
Local embededdness	Our satisfaction about local embeddedness of our supplier/buyer in our important business relationship	Likert scale: total disagree..., total agree	155	3.10	0.80
Equal power distribution	We are equal partners in this business relation	Likert scale: total disagree..., total agree	155	3.12	0.81
Key people leaving	When key person leave our organization/firm, this relationship not continue in future	Likert scale: total disagree..., total agree	155	3.77	1.02
Collaboration with partners	Our collaboration with buyer/supplier in the past	Rating scale: very poor,.....,very good	155	3.26	0.80
Satisfaction with partners	Our satisfaction with buyer/supplier	Rating scale: very poor,.....,very good	155	3.17	0.86
Trust in partner	Trust in our supplier/buyer in our business relationship	Rating scale: very poor,.....,very good	155	3.03	0.77
Commitment with partners	Our commitment with supplier/buyer	Rating scale: very poor,.....,very good	155	2.84	0.76

LIMITATIONS

The questionnaires applied to farmer and producer groups, due to reduced samples, the results cannot be generalised across the country, but they offer significant information regarding the contracting modalities and farmers' contracting choices. The empirical analysis takes into consideration farmers' answers regarding their contracting choice. A small geographical area focus for this study was also one of the limitations.

ACKNOWLEDGEMENT

We wish to thank all the organic Mango farmers, subject matter experts and retailers who patiently shared their time, insights, and views about the organic fruits and vegetable farming and marketing system. Special thanks to Professor Wolfgang Bokelmann, and Dr M. G. Chandrakant, Director, Institute for Social and Economic Change, Bangalore for providing critical comments and inputs on the article. Finally, we extend our special thanks to colleagues at the Humboldt University of Berlin, Germany, for providing us important input during data analysis and in writing the research article.

REFERENCES

Alboiu, C. (2012). Governance and Contractual Structure in the Vegetable Supply Chain in Romania. Romanian Journal of Economic Forecasting, 15(3), 68-82.

Anderson, E., & Weitz, B. (1989). Determinants of continuity in conventional industrial channel dyads. Marketing science, 8(4), 310-323.

Anderson, J. C., & Gerbing, D. W. (1988). Structural equation modeling in practice: A review and recommended two-step approach. Psychological bulletin, 103(3), 411.

APEDA. (2014). National Programme for Organic Production.

Bejou, D., Wray, B., & Ingram, T. N. (1996). Determinants of relationship quality: an artificial neural network analysis. Journal of Business Research, 36(2), 137-143.

Bende, M.J. (2013). Agricultural Profile of Karnataka State. Agricultural Development and Rural Transformation Centre Institute for Social and Economic Change, Bangalore.   A   from http://www.isec.ac.in/Agri%20Profile-Karnataka.pdf. Accessed on 23.11.2016

Boger, S. (2001). Quality and contractual choice: a transaction cost approach to the Polish hog market. European Review of Agricultural Economics, 28(3), 241-262.

Bollen, K. A. (1998). Structural equation models: Wiley Online Library.

Bredahl, L. 2004. Cue utilisation and quality perception with regard to branded beef. Food Quality and Preference, 15 (1), 65-75.

Briz, T., & Ward, R. 2009. Consumer awareness of organic products in Spain: An application of multinominal logit models. Food Policy, 34(3), 295–304.

Bullington, K. E., & Bullington, S. F. (2005). Stronger supply chain relationships: learning from research on strong families. Supply Chain Management: An International Journal, 10(3), 192-197.

Cameron, A. C., & Trivedi, P. K. (2005). Microeconometrics: methods and applications: Cambridge university press.

Cramer, J. S. (2003). Logit models from economics and other fields: Cambridge University Press. New York, USA.

Dimitri C, Oberholzer L, Wittenberger M. (2010). The Role of Contracts in the Organic Supply chain: 2004-2007. United States Department of Agriculture, Economic Information Bulletin 69.

Deliya, M. M. M., Thakor, M. C., & Parmar, B. A Study on "differentiator in Marketing of fresh fruits and Vegetables from Supply Chain Management Perspective".

Dev, S. M. (2012). Small farmers in India: Challenges and opportunities. Indira Gandhi Institute of Development Research, Mumbai.

Devakumar, N. (2014). Organic farming Stake Holders Directory of Karnataka. Book published by Karnataka State Department of Agriculture.

Dyer, J. H., & Singh, H. (1998). The relational view: cooperative strategy and sources of interorganizational competitive advantage. Academy of management review, 23(4), 660-679.

FAO. (2013). Organic supply chains for small farmer income generation in developing countries – Case studies in India, Thailand, Brazil, Hungary and Africa. Agribusiness and Food Industries.

Fischer, C., Hartmann, M., Reynolds, N., Leat, P., Revoredo-Giha, C., Henchion, M., . . . Gracia, A. (2010). Factors influencing contractual choice and sustainable relationships in European agri-food supply chains. European Review of Agricultural Economics, jbp041.

Fischer, C., Hartmann, M., Reynolds, N., Leat, P., Revoredo-Giha, C., Henchion, M., & Gracia, A. (2008). Agri-food chain relationships in Europe-Empirical evidence and implications for sector competitveness. Paper presented at the 12th Congress of the European Association of Agricultural Economists (EAAE), Mathijs E., Verbeke W., de Frahan BH (edts.), Ghent, Belgium.

Fritz, M., & Fischer, C. (2007). The role of trust in European food chains: theory and empirical findings. International Food and Agribusiness Management Review, 10(2), 141-163.

Gliem, J. A., & Gliem, R. R. (2003). Calculating, interpreting, and reporting Cronbach's alpha reliability coefficient for Likert-type scales.

GOI. (2013). Fruits and vegetables Supply Chain in India Ministry of agriculture &Indian institute of foreign trade, 1-5.

Gracia, A., & Albisu, L. M. (2010). Determinants of sustainable agri-food chain relationships in Europe. Agri-food chain relationships, 119.

Gracia, A., de Magistris, T., Albisu, L. M., Fischer, C., & Hartmann, M. (2010). Inter-organizational relationships as determinants for competitiveness in the agri-food sector: The Spanish wheat-to-bread chain. Agri-food chain relationships, 206-219.

Gracia, A. d. M., Tiziana Albisu, Luis Miguel Fischer, C. Hartmann, M. (2010). Inter-organizational relationships as determinants for competitiveness in the agri-food sector: The Spanish wheat-to-bread chain. Agri-food chain relationships, 206-219.

Greenberg, D., & Graham, M. (2000). Improving communication about new food technologies. Issues in Science and Technology, 16(4), 42-48.

Gyau, A., & Spiller, A. (2008). The impact of supply chains governance structures on the inter-firm relationship performance in agribusiness. ZEMEDELSKA EKONOMIKA-PRAHA-, 54(4), 176.

Hair, J. F., Anderson, R. E., Tatham, R. L., & Black, W. C. (1998). Multivariate analysis. Englewood: Prentice Hall International.

Hair, J., Anderson, R., Tatham, R. and Black, W. (2001). Análisis multivariante. V Edición. Prentice Hall. Iberia, Madrid. .

Hair, J. F. (2009). Multivariate data analysis: A global perspective. Springer Science & Business Media. Berlin, Germany.

Hobbs, J. E. (1996). A transaction cost approach to supply chain management. Supply Chain Management: An International Journal, 1(2), 15-27.

Hobbs, J. E. (2004). Information asymmetry and the role of traceability systems. Agribusiness, 20(4), 397-415.

Hughes, A. (1996). Forging new cultures of food retailer-manufacturer relations. Retailing, consumption and capital: towards the new retail geography, 90-115.

Humphrey, J., & Schmitz, H. (2004). 13. Chain governance and upgrading: taking stock. Local enterprises in the global economy: Issues of governance and upgrading, 349.

Kandori, M. (2008). Repeated games. New Palgrave Dictionary of Economics, 2nd edition, Palgrave Macmillan.

Kottila, M.-R., Maijala, A., & Rönni, P. (2005). The organic food supply chains in relation to information management and the interaction between actors.

Lages, C., Lages, C. R., & Lages, L. F. (2005). The RELQUAL scale: a measure of relationship quality in export market ventures. Journal of Business Research, 58(8), 1040-1048.

Lähdesmäki, M., Viitaharju, L., Suvanto, H., Valkosalo, P., Kurki, S., Rantala, S.H., Hartmann, M., Fischer, C., Reynolds, N. and Bavorova, M., (2009). Key Factors Influencing Economic Relationships and Communication in Finnish Food Chains. https://helda.helsinki.fi/bitstream/handle/10138/24734/Reports49.pdf?sequence=1, accessed on 23.11.2016.

Lai, K.-h., Cheng, T., & Yeung, A. C. (2005). Relationship stability and supplier commitment to quality. International Journal of Production Economics, 96 (3), 397-410.

Lairon, D., Huber, M. (2014). Food quality and possible positive health effects of organic products, In Bellon, S., Penvern, S. (Eds.), Organic Farming, Prototype for Sustainable Agricultures, Springer, 295-312.

Lindgreen, A. (2003). Trust as a valuable strategic variable in the food industry: Different types of trust and their implementation. British Food Journal, 105(6), 310-327.

Lindgreen, A., Trienekens, J., & Vellinga, K. (2004). contemporary marketing practice: a case study of the dutch pork supply chain. Paper presented at the Dynamics in Chains and Networks: Proceedings of the Sixth International Conference on Chain and Network Management in Agribusiness and the Food Industry (Ede, 27-28 May 2004).

Lofstedt, R. E. (2006). How can we make food risk communication better: where are we and where are we going? Journal of Risk research, 9(8), 869-890.

Ménard, C., & Valceschini, E. (2005). New institutions for governing the agri-food industry. European Review of Agricultural Economics, 32 (3), 421-440.

Mohr, J. J., Fisher, R. J., & Nevin, J. R. (1996). Collaborative communication in interfirm relationships: moderating effects of integration and control. The Journal of Marketing, 103-115.

Morrissey, W. J., & Pittaway, L. (2006). Buyer-Supplier Relationships in Small Firms The Use of Social Factors to Manage Relationships. International Small Business Journal, 24(3), 272-298.

Mukherjee, W. (2013). Organic food fails to move cash registers for retailers. The Times of India Report(Feb 11, 2013).

Nadvi, K., & Waltring, F. (2004). 3. Making sense of global standards. Local enterprises in the global economy: Issues of governance and upgrading, 53.

Nandi, R., Bokelmann, W., Gowdru, N.V. and Dias, G., 2016. Consumer Motives and Purchase Preferences for Organic Food Products: Empirical Evidence From a Consumer Survey in Bangalore, South India. Journal of International Food & Agribusiness Marketing, 28 (1), pp.74-99.

Naspetti, S. and Zanoli, R., (2009). Organic food quality and safety perception throughout Europe. Journal of Food Products Marketing, 15 (3), pp.249-266.

Naspetti, S., Lampkin, N., Nicolas, P., Stolze, M., & Zanoli, R. (2011). Organic supply chains collaboration: a case study in eight EU countries. Journal of Food Products Marketing, 17 (2-3), 141-162.

Osswald, N. (2010). Sustainable Consumption and Urban Lifestyles: The Case of Hyderabad/ India. Emerging Megacities Discussion Papers 03/2010. Berlin: Europäischer Hochschulverlag.

Osswald, N., & Menon, M. K. (2013). Organic Food Marketing in Urban Centres of India. Bangalore, ICCOA.

Parsons, A. L. (2002). What Determines Buyer-Seller Relationship Quality? An Investigation from the Buyer's Perspective. Journal of Supply Chain Management, 38(1), 4-12.

Rao, V. K. (2006). Market for organic foods in India: consumer perceptions and market potential, findings of a nationwide survey: International Competence Centre for Organic Agriculture.

Ravi, N. (2014). Consumer Preferences and Influencing Factors for Purchase Places of Organic Food Products: Empirical Evidence from South India Indian Journal of Marketing (Vol. 44, pp. 5-17).

Raynaud, E., Sauvee, L., & Valceschini, E. (2005). Alignment between quality enforcement devices and governance structures in the agro-food vertical chains. Journal of Management & Governance, 9(1), 47-77.

Raynolds, L. T. (2004). The globalisation of organic agro-food networks. World Development, 32(5), 725-743.

Raynaud, E., Sauvee, L., & Valceschini, E. (2005). Alignment between quality enforcement devices and governance structures in the agro-food vertical chains. Journal of Management & Governance, 9 (1), 47-77.

Reddy, G., Murthy, M., & Meena, P. (2010). Value chains and retailing of fresh vegetables and fruits, Andhra Pradesh. Agricultural Economics Research Review, 23(2010).

Revoredo-Giha, C., Leat, P., Fischer, C., & Hartmann, M. (2010). Enhancing the integration of agri-food chains: Challenges for UK malting barley. Agri-food chain relationships, 135-149.

Rodríguez, C. M., & Wilson, D. T. (2002). Relationship bonding and trust as a foundation for commitment in US-Mexican strategic alliances: A structural equation modeling approach. Journal of International Marketing, 10(4), 53-76.

Santacoloma, P. (2012). linking smallholders to organic supply chains: what is needed? The International journal for rural development.

Schiemann, M. (2007). Inter-enterprise relations in selected economic activities. Statistics in focus–industry, trade and services, 57.

Schul, P. L., Little, T. E., & Pride, W. M. (1985). Channel climate: Its impact on channel members' satisfaction. Journal of retailing.

Schulze, B., Wocken, C., & Spiller, A. (2006). Relationship quality in agri-food chains: Supplier management in the German pork and dairy sector. Journal on Chain and Network Science, 6(1), 55-68.

Sharma, V., Giri, S., & Rai, S. S. (2013). Supply Chain Management of Rice in India: A Rice Processing Company's Perspective. International Journal of Managing Value and Supply Chains, 4(1), 25-36.

Simatupang, T. M., & Sridharan, R. (2007). The architecture of supply chain collaboration. International Journal of Value Chain Management, 1(3), 304-323.

Stolze, M., Bahrdt, K., Bteich, M.-R., Lampkin, N., Naspetti, S., Nicholas, P., & Zanoli, R. (2007). Strategies to improve quality and safety and reduce costs along the food supply chain.

Talamini, E., Révillion, J. P. (2016). Scale of consumer loyalty for organic food, British Food Journal, 118 (3), 697-713.

Willer, H., Lernoud, J., & Kilcher, L. (2013). The World of Organic Agriculture: Statistics and Emerging Trends 2013: Frick. Switzerland: Research Institute of Organic Agriculture (FiBL) & Bonn: International Federation of Organic Agriculture Movements (IFOAM).

Williamson, O. E. (1991). Comparative economic organization: The analysis of discrete structural alternatives. Administrative science quarterly, 269-296.

Williamson, O. E. (2000). The new institutional economics: taking stock, looking ahead. Journal of economic literature, 595-613.

Winter, M. (2003). Embeddedness, the new food economy and defensive localism. Journal of Rural Studies, 19(1), 23-32.

Young, L. M., & Hobbs, J. E. (2002). Vertical linkages in agri-food supply chains: changing roles for producers, commodity groups, and government policy. Review of Agricultural Economics, 24(2), 428-441.

Zanoli, R., Seljåsen, R., Kristensen, H. L., Kretzschmar, U., Birlouez-Aragon, I., Paoletti, F., Lauridsen, C., Wyss, G.S. Busscher, N., Mengheri, E., Sinesio, F., Vairo, D., Beck, A., Kahl, J. 2016. How to understand the complexity of product quality and the challenges in diffe-rentiating between organically and conventionally grown products – exemplified by fresh and heat-processed carrots (Daucus carota L.), Organic Agriculture, 6 (1), 31-47.

11

Analysis of value chain of sweet potato in two districts of Bangladesh

[1*]S.S.R.M. Mahe Alam Sorwar, [2]Md. Tanvir Ahmed, [3]Sudhir Chandra Nath, [4]Md. Harun -OR- Rashid, [5]Chris Wheatley

[1*,2,3,4] Seed and Agro Enterprise, BRAC, Bangladesh
[5] International Potato Centre, Philippines

Sweet potato plays a significant role in increasing food security and income for the poor farmers of Bangladesh. Sweet potato is mostly grown in the marginal lands of Bangladesh during the period of October to February. It is consumed in different forms e.g. boiled, fries and roasted. Sometimes it is also eaten as a vegetable in curry. The value chain of sweet potato is not well organized in Bangladesh. This study was carried out to analyze the existing value chain of sweet potato in two selected districts of Bangladesh. Quota sampling technique was used to select the samples and primary data were collected through Individual Interview (II), Key Informant Interview (KII) and Focus Group Discussion (FGD) by using structured, open and close ended Questionnaires and check list. Simple descriptive statistics were used to analysis the data. Core value chain actors in sweet potato value chain are input seller, farmers, local trader, retailer and consumers. Mostly farmer cultivates local variety of sweet potato and get a profit around BDT (Bangladesh currency) 30,000 per acre of land. Local trader collects sweet potato both from farmer's field and local market. There are no fixed traders or retailers of sweet potato in the study area. They mostly sell sweet potato along with other vegetables in both urban and local big market. Analysis found that both the trader and retailer gets BDT 3 profit margin by selling 1 kg of sweet potato. No sweet potato processing company was found in Bangladesh though there are huge possibilities and potentials of it in both rural and urban market.

Keywords: Value chain analysis, profit margin, sweet potato, Bangladesh

INTRODUCTION

Sweet potato (*Ipomoea batatas Poir*) is one of the important root crops in Bangladesh as well as in the world. Sweet potato ranks as the world's seventh most important food crop after wheat, rice, maize, potato, barley, and cassava. According to the Food and Agriculture Organization (FAO, 2012), sweet potato was under cultivation in 82,40,969 hectares of land in the world. Considering top 20 sweet potato producing countries in 2012, world's total production was 101,839,463 tons and a majority of which came from China, with a production of 77,375,000 tons (FAO, 2012). Bangladesh produces different varieties of sweet potato but some of the varieties are produced in more quantity based on consumer demand and easy cultivation technique as well as vine (planting materials) availability. It is cultivated more or less in all the districts of the country. However, the suitable areas of sweet potato cultivation are "char land" located on the both sides of rivers.

Corresponding Author: S.S.R.M. Mahe Alam Sorwar, Seed and Agro Enterprise, BRAC, Bangladesh. Email: dilsorwar@gmail.com, mahe.sorwar@brac.net

Table 1. Sampling frame for sweet potato value chain study

Criteria	Location		Total
	Jamalpur	Netrokona	
Sweet potato farmers	50	50	100
Input Seller (Vine, Fertilizer, Pesticide etc.)	4	4	8
Sweet potato Retailer	7	14	21
Sweet potato Trader	8	14	22
Sweet potato Consumer	50	50	100
Key Informant (KI)	2	2	4
Focus Group Discussion (FGD)	1	1	2

In Bangladesh sweet potato is the 4th most important source of carbohydrate after rice, wheat and potato. Sweet potato plays a significant role in increasing food security and income for the poor farmers of Bangladesh (Ahmed *et al.* 2015). The area and production under sweet potato was 24,567 hectare and 297,539 tons, respectively in Bangladesh during the year 2011 (FAOSTAT, 2011). The average per hectare production of sweet potato in Bangladesh is 9.8 tons (FAOSTAT, 2011). Most of the sweet potato producers in Bangladesh are smallholder. Smallholder farmers struggle because of their limited access to inputs (e.g. credit, technology, information) while working on low-productivity land located far distances from output markets via an inadequate, high-cost road system (Lunna and Wilson, 2015). The production of sweet potato has decreased from 435,000 MT in 1995-96 to 253,000 MT in 2011-12 (BBS, 2012).

The value chain describes the full range of activities which are required to bring a product or service from conception, through the different phases of production (involving a combination of physical transformation and the input of various producer services), delivery to final consumers, and final disposal after use (Kaplinsky, 2001). Value chain analysis is a powerful tool to identify the key activities and actors within the whole supply system which form the value chain for that product.

The literature on sweet potato value chain in Bangladesh is very scarce. There are hardly any study found that analysis the value chain of sweet potato. Few studies have found (Ahmed *et al.* 2015; Begum *et al.* 2011) that only deals with the profitability of sweet potato cultivation in Bangladesh. There are various market actors involved from the production to final consumption of sweet potato. Analyzing the roles and activities of these actors has become an important issue to the policy maker in taking decision to improve the overall value chain of sweet potato. Keeping all this issues in mind, the present study was undertaken to map and understand the sweet potato value chain and linkages among different actors in the chain with the intent to identify the gaps and opportunities of strategic interventions to develop the value chains in Bangladesh.

METHODOLOGY

The study was conducted at two selected districts of Bangladesh named Jamalpur and Netrokona. Study areas were selected purposively based on the higher volume of sweet potato production. Samples were taken from Shorishabari and Netrokona Sador sub-district of Jamalpur and Netrokona district, respectively by quota sampling method. Quota sampling is a non-probability sampling technique where the researcher finds and interviews a prescribed number of people in each of several categories. The study team randomly selected samples as input seller (vine, pesticide and micronutrient), producer, traders, retailer, Government officials (Department of Agricultural Extension officers, Bangladesh Agricultural Research Institution etc.). Below the samples covered under this survey, using different tools are shown in tabular format (Table 1).

Primary data were collected through Individual Interview (II), Key Informant Interview (KII) and focus Group Discussion (FGD) by using structured, open and close ended Questionnaires and check list. Sweet potato farmers were directly interviewed by enumerators to collect the primary data on sweet potato cultivation and yield. Secondary data were collected from different publications as well from DAE (Department of Agricultural Extension). Mostly tabular analysis was conducted along with calculating average and percentage. Profit margin of sweet potato farmers, traders and retailer was calculated by using the following formula;

$$NP = TR - TC \quad \dots\dots\dots\dots\dots\dots (1)$$
Where,
NP = Net Profit (BDT)
TR = Total Return (BDT) and
TC = Total Cost (BDT)

RESULTS AND DISCUSSION

The core value chain actors

A wide range of market actors present along the sweet potato value chain. The sweet potato value chain actors found in the study areas were input seller, sweet potato producer, sweet potato trader, sweet potato retailer, sweet potato processor and the consumer.

Availability of inputs

Most of the selected farmers in the study area used local variety of sweet potato for vine multiplication. The study revealed that first time they multiply the vines at their homestead areas and then they use more land to multiply in large scale. There was no nursery found for sweet potato vine multiplication. Farmers were also unaware about the hybrid varieties of sweet potato in the study areas.

The farmers use fertilizer, pesticides (for sweet potato weevil, rootworms, wireworms, white grubs, white fringed beetles etc.) and micronutrient. However, the economic condition of Bangladesh farmers often does not support them to use required quantity of fertilizers due to its high cost (Ali et al. 2009). Most of them don't use the inputs in proper dose and they use very small amount. For that reason the production falls down. Ahmed et al. (2015) found that all the inputs used by the farmers in producing sweet potato are underutilized in Bangladesh.

The labors are not in plenty in the study area. Therefore, during the peak season of weeding and harvesting farmers face shortage of labor. Capital is the scarcest input of all. However, farmers somehow manage it though at a higher interest rate from different micro-finance institutes or local money lenders.

Sweet Potato Producers

The sweet potato producers are defined as the commercial cultivators of sweet potato who later consume a little portion of it and sell the rest. The result found that sweet potato is readily sold to the market (to small traders, large traders, and or trader groups) after three to four month of cultivation. They sale to trader almost 84% of their production whereas 10 % are used for home consumption and 6% are used for livestock feed. Although the production area of sweet potato is decreasing, farmers mostly stated that sweet potato requires few inputs and returns are comparatively high.

As a result, some new farmers are showing interest in cultivating sweet potato in the study area. According to Mendoza (1995) when there are several participants in the marketing chain, the marketing margin is calculated by finding the price variations at different segments and by comparing them with the final price to the consumer. The consumer price is then the base or the common denominator for all marketing margins. Study result found that average production cost of sweet potato is BDT 5.5 per kg while the average farm gate price of sweet potato is BDT 7 per kg. Analysis also showed that sweet potato producer got profit of BDT 29,797 per acre of land (Table 2).

Table 2. Economics of sweet potato at farmers' end

Particular	Amount (BDT)
Per Kg cost of production of sweet potato	5
Per kg farm gate price of sweet potato	7
Total cost of production per acre	35,978
Yield (kg)	9,425
Total value	65,775
Profit per acre	29,797

*1 USD = 80 BDT

Sweet Potato Traders

As sweet potato is very much seasonal crop, the study did not find any specialized traders for sweet potato trading in the study area. The available traders found during the study are basically the seller of seasonal vegetables and horticultural crops. During sweet potato harvesting season they come from different locations to purchase sweet potato both from the farmer's field and local market. They mostly sell sweet potato along with other vegetables in both urban and local market. The study found that, big traders like wholesaler (locally known as Arotdar) do not trade sweet potato as the trading volume is low. There is a variation in offering price to farmer which is based on variety of sweet potato and availability. Sweet potato prices have risen overall and fluctuate depending on the time of the year (Bergh et al. 2012). Analysis showed that at the beginning of the season trader secured 21% profit margin where as in the peak and late season they got 27% and 18% profit margin, respectively (Table 3).

Table 3. Profit margin of sweet potato traders at different period of harvesting season.

Variety	Beginning of the season		Pick of the season		End of the season	
	Purchase Price	Selling price	Purchase Price	Selling price	Purchase Price	Selling price
Local (Red Skin and white flashed)	10	12	9	10	12	15
Profit Margin (Percentage)	17 percent		10 percent		20 percent	
Local (White Skin and white flashed)	11	14	7	8	8	11
Profit Margin (Percentage)	21 percent		13 percent		27 percent	
OFSP Variety	11	14	8	11	9	11
Profit Margin (Percentage)	21 percent		27 percent		18 percent	

Table 4. Profit margin of sweet potato retailers at different period of harvesting season

Variety	Beginning of the season		Peak of the season		End of the season	
	Purchase Price	Selling price	Purchase Price	Selling price	Purchase Price	Selling price
Local (Red Skin and white flashed)	15	20	10	13	15	18
Profit Margin (Percentage)	25 percent		23 percent		17 percent	
Local (White Skin and white flashed)	11	14	8	10	9	11
Profit Margin (Percentage)	21 percent		20 percent		18 percent	
OFSP Variety	12	14	10	13	8	10
Profit Margin (Percentage)	14 percent		23 percent		20 percent	

Most of the retailer sell sweet potato on cash to consumer. Price has been found increasing for last two year, 11% and 18 % respectively in 2011-12 and 2012-13.

Sweet Potato Retailers

Sweet potato retailers are small entrepreneur who are mainly vegetable retailers. They buy in bulk amount (100-110 kg) from traders and almost all the quantity sell to consumer on daily basis. The retailers of the sweet potato in study areas buy from mainly large traders; sometime they directly purchase from field of farmer. Sometimes farmer themselves play a role of retailer in different "hat" (market) day. Result found that sweet potato retailer got 14% profit margin at the beginning of the season where as they secured 23% and 20% profit margin during the peak and late season, respectively (Table 4).

Sweet Potato Processors

By studying the secondary information, it is exposed that sweet potato processing industries are still not available in Bangladesh. Some organizations and International Non-Government Organization (INGOs) are trying to made different food meals for alternate use of sweet potato. Whether in different country, Sweet potatoes are also processed industrially into fried snacks like sweet potato fries (chips), candy, starch, noodles, and flour (Bergh at. al 2012).

Sweet Potato consumers

Study found that consumer prefer to purchase local (red skin and white flashed) variety of sweet potato because of its taste and traditional practice. Majority of the consumers prefer boiled form of sweet potato followed by burned form. In some areas sweet potato is also used as curry. Daily intake (on an average) of sweet potato from April to June was found to be 299 gm per person at Jamalpur and 166 gm per person at Netrokona. Usually they buy sweet potato from local market. All the respondents at Netrokona and Jamalpur mentioned that their family members also like to eat sweet potato.

Eighty Five percent of the consumers of Netrokona and 76 percent from Jamalpur reported that they know about the nutritional values of sweet potato. The entire sampled

consumer knows that sweet potato leaves can be eaten as leafy vegetable and they love to eat sweet potato leaves. The leaves are collected (small scale) from farmer's own land for home consumption.

Storage

The result of this study found that only nine percent farmers stated that they store sweet potato at home for 1-3 months. Core problem of this storing is insect and fungus attack. Farmers and traders stated that they don't usually store as they don't trade sweet potato in the off season. However survey revealed that about 90 percent farmers and 45 percent traders are interested to store sweet potato in a cheap and healthy way without any post harvest loss.

The supporting function players

The supporting function players for the sweet potato value chain are those who are not directly related to the sweet potato value chain but provide different supports to the value chain actors. The support functions include different services (e.g. credit), research and development, infrastructure, and information. In the study area it was found that there are several NGOs are operating who offers credit support to the farmers. Moreover there are also specialized agricultural bank presents in the study area.

Information service providers

In order to provide information to the farmer, Government of Bangladesh have established DAE (Department of Agricultural Extension) who communicate with the farmers about cultivation process, pesticide use, seed use, sale of improved quality seed etc. DAE officials like SAAO (Sub Assistant Agricultural Officers) carry on such tasks but they are very limited in number compared to the number of farmers in need of guidance. Sometimes, input sellers while selling fertilizer, pesticide and seed are informing farmers but still these input sellers do not have that much capacity to provide services demanded by the farmers.

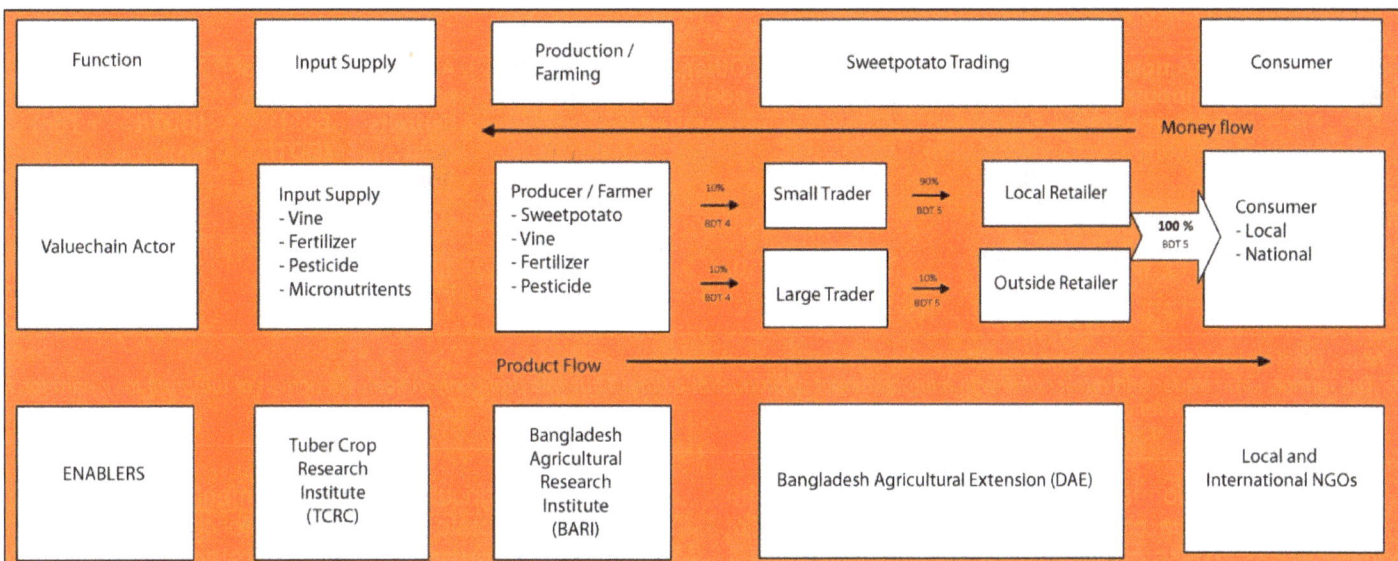

Figure 1. Value Chain Map of Sweet potato in two Districts of Bangladesh

Table 5. Channels in the sweet potato value chain in the study areas

Channel Numbers	Input Supplier	Farmers	Traders	Large Traders	Retailer (Local)	Retailer (outside)	Consumers (Local)	Consumers (National)	percentage
1	■	■	■		■		■		40
2		■			■		■		10
3	■	■				■		■	40
4	■				■				10

Financial service providers

Different financial service providers like NGOs, MFIs and specialized agricultural bank etc. are active at field level for financial service. But the sweet potato farmers are not receiving sufficient service regarding finance related issue. Farmers are seen borrowing money from neighbours at the time of cultivation.

The sub-sector map

A subsector is defined as a vertical grouping of enterprises involved in the production and marketing of one well-defined product or several closely related products (Boomgard et al., 1992). According to ILO (2009), mapping a chain means creating a visual representation of the connections between businesses in value chains as well as other market players. The value chain map is a graphical presentation of the value chain actors and other players. Here the regulators, standard setters, law or policy makers, informal rules and norms setters are shown at the top portion. In the middle are the value chain actors – from left to right, at the bottom are the support function players.

Channels in the sweet potato value chain

According to Lundy *et al.* (2004) a market chain is used to describe the numerous links that connect all the actors and transactions involved in the movement of agricultural goods from the farm to the consumer. The study had found very few channels for sweet potato; they are mentioned below both in written and graphical form. It is mention worthy here that consumers are not part of any channel of value chain since they do not add any value. However, they have been shown here only for the sake of clarifying the flow of final product to the ultimate hand.

Channel 1: Input Suppliers – Farmers - Traders – Retailers (Local) - Consumers
Channel 2: Input Suppliers – Farmers - Traders – Large Trader – Retailers (Outside) - Consumers(National)
Channel 3: Input Suppliers – Farmers - Retailers (Local) - Consumers
Channel 4: Farmers - Retailers (Local) – Consumers

Cost of different sweet potato value chain actors

In Table 6, the cost of farmers, traders, and retailers are

Table 6. Monetary flow of sweet potato in different level

Sweet potato Actors	Amount of Inputs Purchased (ton)	Price of Inputs Purchased (per unit) BDT	Cost of Purchased Amount BDT	Other cost of operation BDT	Total Cost of Operation (BDT)	Amount of Outputs Sold (ton)	Price of Outputs Sold (BDT/ unit)	Total Sales (BDT/t on)	Income (BDT/pe r Ton)
Farmers					3,000	1	7,000	7000	4,000
Traders	1	7,000	7,000	0	7,000	1	12,000	12000	5,000
Retailers	1	12,000	12,000	0	12,000	1	17,000	17000	5,000

Note: The wastage (during caring and transport) in the different level is negligible.
*The farmer, large trader and retailer deal with a large amount of product at a time. Thus this profit only indicates a portion of the profit they earn for 1 ton of sweet potato, not for one transaction.

shown in tabular format to show the monetary flow in the sweet potato value chain. Main value chain consists of these actors. These calculations are shown in terms of 1 ton (1000 kg) sweet potato production and distribution in a year.

CONCLUSION AND RECOMMENDATION

Though sweet potato plays an important role in increasing income and food security of the farmers but, the value chain system was not strength enough in the study area to get the potential benefits out that. Sweet potato is produced and consumed locally while there is demand for it in the capital market of the country. Due to unstructured market management it has not been reckoned as lucrative product for investment. Lack of supply demand synchronization and limited availability and demand for value added products from sweet potato and a lack of knowledge of diversified products are the key drawback to strengthen the existing sweet potato value chain in Bangladesh. The study suggests arranging the linkage between the traders of the capital market and the local traders of the study areas. From the study it has been revealed that there are numbers of constraints exist both at upstream and downstream level of sweet potato value chain. Majority of the farmers are not aware about the food value of sweet potato. At the grower level roughly 14 percent farmers know about the improved varieties of sweet potato. Though 80 percent farmers use fertilizer specially Urea, TSP and MoP but they use it in small proportionate than the recommended dose due to their poor economical condition and mostly for careless farming. Getting quality vines is also a big problem for the farmers. By addressing proper management of crop cultivation, sweet potato farmer can get higher yield and that can add more profit to them.

ACKNOWLEDGEMENT

The Author is very much grateful to BRAC, International Fund for Agricultural Development (IFAD) and International Potato Centre (CIP) for granting fund to undertake this research. The authors acknowledged the contributions of Dr. Lighton Dube, Dr. Mohammad Jabbar, Dr. Claver Ngaboyisonga, Sendhil R, Dr. John Walsh, Dr. Kassa T. Alemu, Dr. Giuseppe Di Vita, Dr. Maziku Petro, Dr. Abel Mafukata and Dr Muhammad Kadwa for donating their time, critical evaluation, constructive comments, and invaluable assistance toward the improvement of this very manuscript.

REFERENCES

Ahmed MT, Nath SC, Sorwar MA, Rashid MH (2015). Cost-effectiveness and resource use efficiency of sweet potato in Bangladesh. Journal of Agricultural Economics and Rural Development, volume 2 (2), pp. 26-31.

Ali MR, Costa DJ, Abedin MJ, Basak AC (2009) Effect of fertilizer and variety on the yield of sweetpotato. Bangladesh J. Agril. Res. 34(3): 473-480.

BBS (2012). Statistical Year Book of Bangladesh. *Bangladesh Bureau of Statistics Division, Ministry of Planning, Government of the People Republic of Bangladesh, Dhaka, Bangladesh.*

Begum MEA, Islam MN, Alam QM, Hossain SB (2011). Profitability Of Some Bari Released Crop Varieties In Some Locations Of Bangladesh. *Bangladesh Journal of Agricultural Research*, 36(1), 111-122.

Bergh K, Orozco P, Gugerty MK, Anderson CL (2012). Sweet Potato Value Chain: Nigeria. Weven School of Public Affairs. EPAR Brief No. 220

Boomgard JJ, Davies SP, Haggblade SJ, Mead DC (1992). A subsector approach to small enterprise promotion and research. *World Development*, 20(2), 199-212

FAO (2012): Food and Agricultural Organization. http://faostat.fao.org/site/567/DesktopDefault.aspx?Pagel D=567#ancor

FAOSTAT (2011). Food and Agricultural Organization http://faostat.fao.org/site/567/DesktopDefault.aspx?Pagel

http://faostat.fao.org/site/567/DesktopDefault.aspx?P ageID=567#ancor

ILO (2009). Value chain development for decent work: A guide for development practitioners, government and private sector initiatives. [http://www.ilo.org/empent/Publications/WCMS_1154 90/lang--en/index.htm] site visited on 25/12/2012.

Kaplinsky R, Morris M (2001). *A handbook for value chain research* (Vol. 113). Ottawa: IDRC

Lunaa F and Wilson PN (2015). An Economic Exploration of Smallholder Value Chains: Coffee Transactions in Chiapas, Mexico, *International Food and Agribusiness Management Review, 18*(3).

Lundy M, Gottret MV, Cifuentes W, Ostertag GCF, Best R, Peters D, Ferris S (2004). *Increasing the competitiveness of market chains for smallholder producers: Module 3: Territorial approach to rural-agroenterprise development.* CIAT.

Mendoza, G (1995). A Primer on Marketing Channels and Margins. Prices, Products and People: Analyzing Agricultural Markets in Developing Countries. Lynne Reinner Publishers, London, United Kingdom. 498pp.

The role of ICT in facilitating farmers' accessibility to extension services and marketing of agricultural produce: The case of Maize in Mbozi District, Tanzania

Francis Lwesya[1*] and Vicent Kibambila[2]

[1,2] Department of Business Administration, School of Business Studies and Economics, University of Dodoma, Tanzania.
* Corresponding author's email: flwesya@yahoo.com

The rapid pace of ICT development and its consequent use across economic, social and political spectrums has raised concerns among policy makers and practitioners over its potential to spur productivity in the agriculture sector as well. This paper examines the role that ICT can play in facilitating smallholder farmers' accessibility to extension services and marketing of agricultural produce in Mbozi District. The study used structured questionnaires to collect information. A sample of 250 farmers was selected randomly and interviewed.The findings reveal that farmers are using ICT facilities to get access to extension services and in marketing maize in Mbozi District. The most preferred and major ICT tools used to inquire and receive extension services and market information are the mobile phones (53.88%), radio (23.67%), television (14.69%) and the internet (7.75%). However, effective use of ICT in the study area is constrained by poor infrastructure in rural areas, and lack of technical know-how exhibited in the lack of basic ICT skills. Other constraints are the erratic power supply, poor signals, lack of network, and lack of internet connectivity and a high cost of some ICT tools. This suggests that if requisite ICT infrastructures are put in place in Mbozi District, ICT can bring about significant benefits to smallholder farmers leading to increased agricultural productivity and hence poverty reduction. Thus, the study recommends promoting investment in renewable energy sources in order to address the problem of power in rural areas. Creating an enabling environment for ICT services accessibility, including the construction of transport and communication network infrastructures. The establishment of market information centers and telecenters in Mbozi District particularly in rural areas and advocating for the use of collective marketing through strengthened farmers groups to ease their access to ICT facilities.

Keywords: ICT, agricultural produce, maize, extension services, farmers, market Information

INTRODUCTION

For the past three decades, there has been unprecedented growth in the use of Information and Communication Technology (ICT) in various social and economic fields across the world. Matambalya and Wolf (2001) observed that through the rapid spread of ICT and ever decreasing prices for communication, markets in different parts of the world become more integrated. Moreover, the report by Economic, Social, and Research

Foundation (ESRF) in Tanzania (2009), claims that recent developments in ICT have greatly increased the opportunities for people to "connect" virtually without the absolute need for physical contact for social or trading purposes.

According to Lewis (2009), ICT is an umbrella term that includes any communication device or application, encompassing: radio, television, mobile phones, computers and network hardware and software, satellite systems and so on, as well as the various services and applications associated with them, such as videoconferencing and distance learning. This definition was qualified by Angelo and Wema (2010), who stated that ICT are the techniques, methods and tools used to access information and to communicate with others. It consists of all technical means used to handle information and aid communication, including computer and network hardware as well as necessary software. On the other hand, Birner et al (2009), Christoplos, (2010), Davis and Heemskerk (2012), define agricultural extension and advisory services as systems that facilitate the access of farmers, their organizations, and other value chain and market actors to knowledge, information, and technologies, facilitate their interaction with partners in research, education, agribusiness, and other relevant institutions; and assist them to develop their own technical, organizational, and management skills and practices as well as to improve the management of their agricultural activities. While, USAID (2013) defines agricultural marketing as the process of identifying, communicating with, and maintaining relationships with buyers of a producer's products to directly affect volume, value and timing of sales. Marketing activities enable a producer to find new buyers, build and maintain relationships with current buyers, and access market research to manage supply, anticipate demands and establish prices. Farmers' accessibility to extension services and access to better markets of agricultural produce are important aspects for raising agricultural productivity, ICT seems to offer strategic and cheap solutions in accessing both extension services and markets for agricultural produce as USAID (2013), remarks that ICT solutions can increase efficiencies and improve competitive dynamics in agriculture, which can raise agricultural productivity and incomes and increase food security.

Following the development of ICT across the world, its usage has gained prominence in many developing countries as well. Njelekela and Sanga (2015), state that many activities in the modern world are becoming more dependent on the application of ICTs in one use or another. The benefit of ICTs has increased in such a way that it reaches even those who do not themselves have first-hand access to them. For instance in Tanzania ICT

is being used in sectors like agriculture, education, transport and communication, manufacturing and construction both in public and private domains. This is because of its importance in fostering social and economic development as ESRF (2009), states that the use of ICTs is reducing transaction cost, time, and space barriers, allowing mass production of customized goods and services.

Looking at the agriculture sector, Tanzania has registered some progress in the use of ICT despite its usage still being at an infancy stage. It is believed that an effective use of ICT in agriculture can spearhead significant development in the sector. This is because 70% of the population lives in rural areas and depends on agriculture for their livelihood. ICTs are crucial in facilitating communication and access to information for agricultural and rural development. Since agriculture is the national priority sector, it is one of the potentially beneficial areas for the application of ICTs for economic transformation. Thus, if there is an effective application of ICT in the agriculture sector, farmers could be informed of inputs prices, market information, in that way they will be able to increase the yield of products. Bolarinwa and Oyeyinka (2011), states that there will be a timely exchange of agricultural information between the extension agents and farmers if ICT components are integrated with the delivery of agricultural information to farmers. Different tools of ICT such as email, SMS, website and TV can aid interaction and communication between researchers, extension agents, and farmers, which in turn will help to develop market linkages and producers can become more informed about the needs of consumers. Sutrisno and Lee (2010) state that these technologies are increasingly being seen as cost effective and as practical tools to facilitate information delivery and knowledge sharing among farmers, extension agents, and other stakeholders.

Since to a large extent in rural areas agriculture is carried out by smallholder farmers, they are exposed to many challenges including shortage of reliable markets, lack of reliable sources of market information, falling labor and land productivity due to an application of poor technology and dependence on unreliable and irregular weather conditions. Further, Tadesse and Shively (2013) notes that the village markets are characterized by asymmetric information in which traders are more informed than farmers about the prices in the central or regional markets. In such circumstances, ICT can be critical in assisting smallholder farmers to raise agricultural productivity by addressing market information asymmetry problem and facilitating their accessibility to extension services. According to Winrock (2003), Knowledge, communication, and information exchange have influenced farmers decision on what to plant, when to plant, how to cultivate and harvest, where to store and where to sell and at what price. That is why the World Bank (2011) stated that the proliferation of adaptable and more affordable technologies and devices has also increased ICT's relevance to smallholder agriculture and improvement of agricultural productivity will be realized when farmers are linked to market information.

Maize is one of the major staple foods for the majority of Tanzanians; it provides 60% of dietary calories to more than 37 million Tanzanians. The Southern Highland regions of Iringa, Songwe, Ruvuma and Rukwa are the most important producers of maize, they account for 35.9% of the national output (Sanga, 2013). For the case of Songwe region, Mbozi District is the leading maize producer. The district lies between latitude 8° and 9' South of Equator and longitudes 32° 7' and 33° 2' East of Greenwich Meridian. The district share borders with Mbeya district to its eastern part, Ileje district to the South, Zambia and Rukwa region to the west and Chunya district to the North. Mbozi district is divided into three distinct agro-ecological zones. These are the South Western Plateau zone (Ndalambo plateau), the Coffee zone (Mbozi plateau) and the Lowlands of Lake Zone (Msangano plains). The District lies between 900 and 2750 meters above the sea level. On average it receives rainfall between 1350 mm and 1550 mm per annum; while temperature ranges between 200°C to 280° C. There is a large arable land which is suitable for agriculture, out of 766,640 Ha of arable land only 216,198 Ha is cultivated, which is equivalent to 28% of the total arable land in Mbozi District. The existence of a good number of rivers and fertile valley suitable for irrigation is another sign for the potential of the area in agricultural production. However, Mbozi maize farmers face many challenges such as high transport and transactional costs, small and inefficient markets, low agricultural productivity and slow adoption of new technologies. Nevertheless, according to Tofik (2014), awareness of up-to-date market information on prices for commodities, inputs and consumer trends can improve farmers' livelihoods substantially and have a dramatic impact on their negotiating position.

This paper explores the role ICT has in facilitating farmers' accessibility to extension services and marketing of maize in Mbozi district. The specific objectives of this paper were to determine;

i.   What are the major ICT tools through which farmers get access to agricultural and market information?
ii.   What are the attitudes of smallholder farmers towards ICT?
iii.   What are the major factors that hinder the effective use of ICT?

METHODOLOGY

The study was conducted in Mbozi District, a district which is one of the largest sources of Maize in Songwe region. Prior to actual data collection, a pilot test was conducted in order to establish whether the questionnaires are capturing the intended objectives or not, by looking if the wording is clear, questions are interpreted in the same way by the respondents, and if there is any research bias and what response is provided. Researchers also ensured that the language was clear to assist respondents in answering questions and instructions given clearly. Thus, a pilot study was conducted to 40 respondents in Mbozi District. The objective of the pilot study was to pre-test the research questionnaires and helped researchers to refine the items which were not clear to the respondents. During the actual data collection, a total of 25 villages of Mbozi District were visited. Interviews, desk work review, and structured questionnaires were used to collect information and 250 questionnaires were distributed to randomly selected farmers. A 5–point Likert scale ranging from 1 as strongly disagrees to 5 as strongly agree was used for the measurement. The reliability analysis was used to detect the internal consistency of the farmers' responses using Cronbach's alpha. Cronbach,s alpha is a coefficient (a number between 0 and 1) that is used to rate the internal consistency or the correlation of items in a test. The coefficients for the variables chosen for the study should have to be more than 0.70, to consider it as an acceptable value (Nunally, 1978). The paper uses descriptive statistics to analyze the role of ICT in facilitating marketing of Maize farmers' access to extension services and marketing of Maize in Mbozi District. In order to capture the attitudes of farmers on the role of ICT in Mbozi district, we used the model applied by Aydin and Tasci's (2005) in which the alternatives were designed in a way that provides easy coding and assessment of the users. The alternatives were coded as 1, 2, 3, 4 and 5 as in a five point likert scale. Since this paper uses a five point likert scale measurement, this model fits properly. According to Aydin and Tasci's (2005), the 3.41 mean score can be identified as the expected level of readiness, while other responses enable organizations to show higher or lower levels of readiness. The mean average of 3.4 was determined after identifying the critical level: 4 intervals/5 categories = 0.8 as shown below.

STUDY FINDINGS AND DISCUSSIONS

Characteristics of the sampled respondents

Of the total 250 questionnaires distributed, 245 were collected. 156 (63.67%) were male and 89 (36.33%) were female (table 1). Age wise, 31 to 60 years (40%) a

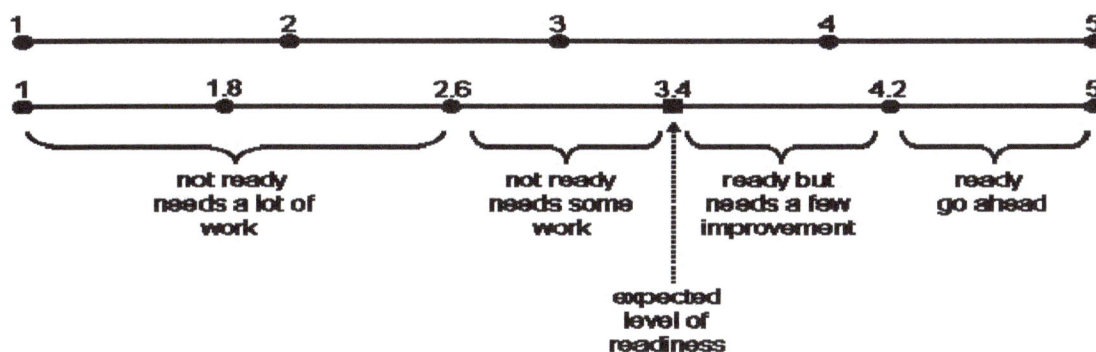

Table 1: Selected Socio-economic Characteristics of Respondents

Item	Frequency	Percentage
Gender		
Male	156	63.67
Female	89	36.33
Total	245	100
Age		
Less than 30 years	44	17.96
31 to 60	98	40
61 to 75	62	25.31
Above 75	41	16.73
Total	245	100
Education		
Informal education	33	13.47
Primary education	98	40
Secondary education	71	28.98
College education	20	8.16
University education	23	9.39
Total	245	100
House hold size		
1 to 15	75	30.61
16 to 25	89	36.33
27 to 30	58	23.67
31 to 45	14	5.71
Above 46	9	3.67
Total	245	100
Monthly Income (Tanzania Shillings)		
less than 100000	18	7.35
101000 to 200000	64	26.12
201000 to 400000	58	23.67
401000 to 600000	50	20.41
601000 to 1000000	25	10.20
Above 1000000	30	12.24

Table 1: Cont.: Selected Socio-economic Characteristics of Respondents

Total	245	100
Years of farming		
1 to 15	79	32.24
16 to 30	104	42.45
31 to 45	42	17.14
above 46	20	8.16
Total	245	100
Area of farming		
1 to 2 Acres	84	34.28
3 to 6 Acres	98	40
7 to 10 Acres	49	20
11 to 15 Acres	14	5.72
Total	245	100

Source: Field Survey (2016)

Table 2: Major ICT tools farmers use to get access to agricultural and market information in Mbozi District

ICT tools used		Frequency	Percentage
Mobile phone	Calling	76	31.020
	SMS	56	22.857
Radio		58	23.673
Television		36	14.69
Internet	Website	10	4.082
	Email	9	3.673
Total		245	100.00

Source: Field Survey (2016)

group with the most respondents, followed by those between 61 to 75 years (25.31%). Respondents less than 30 years (17.96%) and the relatively lower number of respondents were above 75 years (16.73%). This suggests that above 75 is an ageing population, according to Chete and Fasoyiro (2014), an ageing population will likely affect productivity in a negative way and reduce volume of sales or market participation. In terms of the level of education most of the respondents had primary education (40%), followed by those who had secondary education (28.98%). Respondents who had not received any formal education were (13.47%). Those who had University and College education were 9.39% and 8.16% respectively. This suggests that most of the respondents had informal, primary and secondary education. Lack of education or low level of education may mean the majority of farmers could not be able to apply ICT tools effectively in the agriculture sector to their advantage as Matungal et.al (2001) note that a higher level of education is desirable to minimize costs of search and screening information and transaction cost in both factor and product market.

The respondents (36.33%) had the highest household size of between 16 and 25, followed by 1 to 15 (30.61%). The households size 27 to 30 (23.67%) while 31 to 45 and above 46 had 5.71 and 3.67 respectively. According to Lapar et.al, (2003), the propensity to participate in the market economy declines with the number of household members. Most of the farmers had working experience of between 16 to 30 years (42.45), followed by 1 to 15 years (32.24%), then 31 to 45 (17.14%) and lastly above 46 (8.16%). This suggests that most farmers had several years of farming experience, according to Matungal et.al (2001), the expectation is for farmers with higher farming experience to have higher commercialization index and thus better participation in the markets.

Major ICT tools farmers use to get access to agricultural and market information in Mbozi District

The major ICT tools used by farmers in Mbozi district are mobile phone, Radio, television and the Internet. A total of 132 farmers use mobile phones, whereas 31.020% use for calling while 22.86% use text SMS to inquire and

Table 3: Reliability Analysis

Item	Cronbach's Alpha
Attitudes of farmers on the use of ICT	
ICT tools are easily accessible by farmers	0.785
ICT tools are easier for farmers to learn how to use	0.788
ICT tools are easy for farmers to use and operate	0.797
ICT tools are valuable information source for farmers	0.783
ICT tools enhance agricultural productivity	0.785
ICT improves the quality of service offered to farmers	0.782
ICT enables farmers to get correct and updated agricultural and market information	0.775
ICT improves farmers communication with agricultural extension workers and input suppliers	0.761
ICT enables farmers to reach new markets	0.766
ICT increases farmers profitability	0.782
ICT reduces travel time and expense	0.771
ICT leads to improved negotiation power	0.764
ICT leads to broader network	0.776
ICT helps secure better markets and prices	0.773
Factors that hinder the usage of ICT tools	
ICT tools are too expensive	0.762
Language barrier	0.768
Erratic power supply	0.775
Low education/literacy	0.774
lack of internet connectivity	0.765
Poor signals or reception (lack of network)	0.766
lack of technical knowhow i.e. lack of basic ICT skills	0.772
Poor infrastructures in rural areas hinder effective use of ICT	0.779
Lack of ICT related facilities e.g tele-centers	0.789

receive agricultural and market information. With the increase of mobile service subscribers in Tanzania, their services have penetrated even in rural areas. There are approximately 3.18 million core mobile subscribers in the market, compared to 2,963,737 in 2005. Internet service subscribers have increased to about 11,000,000 by December 2014 from 3,563,732 in 2008. Kapange (2012) observes that with the mushrooming of mobile phones in Tanzania are increasingly becoming affordable, and they help overcome rural isolation and make communication easier. The wireless technologies that have entered remote rural areas have reduced reliance on costly fixed telephone infrastructures. This has propelled the use of mobile phones in the agriculture sector in Mbozi District as well. Those who use radios were 23.67%. This is because radios have been one of the traditional means of receiving information for a long time particularly in rural areas due to its affordability and wide reception in the absence of network or electricity. These factors make it one of the most preferred ICT tool. The farmers who use television were 14.69% while 19 farmers use internet, among those 4.08% use website services while 3.67% use email. Internet is the ICT tool with relatively lower number of users in Mbozi District, mostly the knowledgeable and educated farmers were found to be using internet.

Reliability Analysis

The Reliability analysis shows that all the factors have shown alpha value greater than 0.7 (Table 3), suggesting the presence of internal consistency in the instrument of measurement. The overall Cronbach's Alpha is 0.783. The factors and dimensions included for analysis carry a good degree of reliability to support the objectives of the study. Hence it can be concluded that the data collected is highly reliable.

Attitudes of farmers on the use of ICT in Mbozi District

Table 4 presents the farmers attitudes on the role of ICT in marketing of maize and facilitating their access to extension services in Mbozi District. Based on these findings, many farmers seemed to appreciate the importance of ICT in the agriculture sector. For instance farmers stated that ICT tools are valuable information source for farmers. This item had 4.331 (86.62%) mean and the views of the farmers in this component were the most consistent with coefficient of variation of 19.6%. Also item, ICT enhances agricultural productivity had 4.1184 (82.30%) and the views expressed were consistent by 27.3% coefficient of variation (C.V). In

Table 4: Attitudes of farmers on the role of ICT

NO.	Perception statement	Mean	Std. Deviation	Coefficient of variation (C.V) (%)
1	ICT tools are easily accessible by farmers	3.9673	1.07455	27.085
2	ICT tools are easier for farmers to learn how to use	3.8612	0.97779	25.32
3	ICT tools are easy for farmers to use and operate	3.8082	1.11984	29.40
4	ICT tools are valuable information source for farmers	4.3306	0.84983	19.62
5	ICT enhances agricultural productivity	4.1184	1.13363	27.52
6	ICT improves the quality of service offered to farmers	3.5184	1.19282	33.90
7	ICT enables farmers to get correct and updated agricultural and market information	4.2612	0.90825	21.31
8	ICT improves farmers communication with agricultural extension workers and input suppliers	3.8000	0.96920	25.51
9	ICT enables farmers to reach new markets	3.6327	1.06920	29.43
10	ICT increases farmers profitability	4.2327	0.96185	22.72
11	ICT reduces travel time and expense	3.9184	1.04483	26.66
12	ICT leads to improved negotiation power	3.9469	1.08324	27.44
13	ICT leads to broader network	4.0286	0.98097	24.35
14	ICT helps secure better markets and prices	3.9061	1.14312	29.26

Source: Field Survey 2016
Decision rule: the mean above 3: Ready but needs few improvements = Farmers have a positive attitude on ICT

terms of whether ICT enables farmers to get correct and updated agricultural and market information, the item recorded 4.2612 (85.2%) and the views of farmers were consistent by 21.3% coefficient of variation. Also, farmers stated that ICT increases farmers profitability, the mean was 4.26 (84.65%) and 22.7% coefficient of variation. Further, farmers agreed as well that ICT leads to broader network with the mean of 4.0286 (80.6%) and the views expressed were consistent by 24.4% coefficient of variation. However, regarding whether ICT improves farmers' communication with agricultural extension workers and input suppliers, the views expressed were above the ready but needs a few improvement level, the mean was 3.8 (0.76%) and the views expressed were consistent by 25.5% coefficient of variation. The results suggest that the farmers have a positive attitude on the ICT use in Mbozi District. The positive perception of farmers on ICT reflects what was highlighted in the URT (2007) report which stated that ICT has currently spread even to the remote rural area which in the past lagged behind and as such, Obayelu and Ogunlade, (2006) state that due to this drastic change, the farmers in rural areas have been aware of the various issues of interest to them that affect their livelihood. The views are shared also by Chapman and Slaymaker (2002) who argue that any ICT intervention that improves the livelihoods of poor rural families will likely have significant direct and indirect impacts on enhancing agricultural production, marketing and post-harvest activities which in turn can further contribute to poverty reduction.

Factors that hinder the usage of ICT tools in Mbozi District

Table 5 shows the barriers associated with ICT usage in Mbozi District. The major constraints to the use of ICTs were lack of ICT related public facilities 4.17 (83.4%), poor infrastructures in rural areas 4.16 (83.2%), low education/literacy 4.04 (80%) and lack of technical knowhow i.e. lack of basic ICT skills 3.96 (79.2%). These items recorded the highest scores. Erratic power supply 3.95 (79%), language barrier (3.89) and Poor signals or reception (lack of network) 3.87 (77.4%), lack of internet connectivity 3.85 (77%) and that ICT tools are too expensive 3.84 (76.8%) recorded relatively lower scores. This suggests that the identified barriers are a hindrance to the effective use of ICT in Mbozi District since all measurement instruments have recorded its mean above 3. The findings reflect what was observed on the ground where factors like low education/literacy and lack of basic ICT skills resulted in slower adoption of ICT tools to some farmers particularly to informal and primary school education holders. This informs that knowledge of ICT and having basic skills are important aspects if farmers are to use ICT effectively. Basic knowledge and skills can enable farmers to take advantage of going beyond the local market through global supply chains via the Internet, etc. Poor infrastructures in rural areas is another barrier for the effective use of ICT in Mbozi District, this because most remote villages seemed to lack infrastructures to enable proper connectivity with other parts of the District.

Table 5: Factors that hinder the usage of ICT tools

Perception statement	Strongly agree	Agree	Neutral	Disagree	Strongly disagree	Mean
ICT tools are too expensive	66 (26.9%)	99 (40.4%)	57 (23.3%)	20 (8.2%)	3 (1.2%)	3.84
Language barrier	94 (38.4%)	70 (28.6%)	47 (19.2%)	27 (11%)	7 (2.9%)	3.89
Erratic power Supply	86 (35.1%)	104 (42.4%)	28 (11.4%)	10 (4.1%)	17 (6.9%)	3.95
Low education/literacy	78 (31.8%)	122 (49.8%)	29 (11.8%)	9 (3.7%)	7 (2.9%)	4.04
lack of internet connectivity	77 (31.4%)	96 (39.2%)	45 (18.4%)	13 (5.3%)	14 (5.7%	3.85
Poor signals or reception (lack of network)	75 (30.6%)	89 (36.3%)	57 (23.3%)	21 (8.6%)	3 (1.2%)	3.87
Lack of technical knowhow i.e. lack of basic ICT skills	100 (40.8%)	65 (26.5%)	59 (24.1%)	13 (5.3%)	8 (3.33%)	3.96
Poor infrastructures in rural areas hinder effective use of ICT	117 (47.8%)	79 (32.2%)	25 (10.2%)	20 (8.2%)	4 (1.6%)	4.16
Lack of ICT related facilities i.e Tele-centers	118 (48.2%)	79 (32.2%)	24 (9.8%)	20 (8.2%)	4 (1.6%)	4.17
Total mean						27.91
Grand mean						3.99

Source: Field Survey, 2016

Roads and bridges, especially in developing countries form part of the ICT infrastructure. Very limited agricultural products are delivered over the information infrastructure or the Internet. Most of the agricultural products purchased are still delivered the conventional way (physical delivery). Hence, poor roads and bridges, inefficient transport systems, coupled with the high cost of transport are among the obstacles in the uptake of ICT in Mbozi District.

CONCLUSION AND RECOMMENDATIONS

The objective of the study was to examine the role of ICT in facilitating marketing of agricultural produce and farmers' accessibility to extension services in Mbozi District. Based on the attitudes of farmers, the ICT tools mostly used are mobile phones (53.88%), Radio (23.6%), television (14.69%) and the internet (7.75%) to inquire and receive agricultural and market information. While the findings seem to suggest that there is progress towards the use of ICT in the study area and there is an appreciation of the importance of ICT in maize production and marketing, farmers are constrained by poor infrastructures in rural areas, low education/literacy and lack of technical know-how i.e. lack of basic ICT skills. Other constraints are the erratic power supply, language barrier, poor signals or lack of network, lack of internet connectivity and the high cost of ICT tools. This study has the implication that if requisite infrastructures are put in

place in rural areas, particularly where agricultural activities takes place, ICT can bring significant benefits to smallholder farmers by raising agricultural productivity and hence leading to poverty reduction. Therefore, the study recommends that establishing market information centers and telecenters in Mbozi rural areas should be done so as to expand access to market information. Promoting investment in renewable energy sources in order to address the problem of power as they are relatively cheap and sustainable and encouraging more investments in the national physical and transport infrastructure; and creating enabling environment for ICT accessibility, including more construction of infrastructure with particular emphasis on networks which are already widely available (television, radio and mobile phones).

REFERENCES

Angelo C, Wema E (2010). Availability and usage of ICTs and e-resources by livestock researchers in Tanzania: Challenges and ways forward, International Journal of Education and Development using Information and Communication Technology, Vol. 6, Issue 1, pp. 53-65.

Aloyce M (2005). ICT for improved crop marketing in rural Tanzania: Project summary. Retrieved from http://www.uneca.org/aisi/iconnectafrica/v2n2.htm

Aker JC (2008). Does Digital Divide or Provide? The Impact of Mobile phones on Grain Markets in Niger. Center for Global Development Working Paper No. 154.

http://www.cgdev.org/content/publications/detail/89441 0, accessed January 2011.

Annerose D (2010). "ICT for Social and Economic Development." Presentation by Manobi at the World Bank, Washington, DC, August.

Tanzania Agricultural Marketing Policy (2008)

Birner RK, Davis J, Pender E, Nkonya P, Anandajayasekeram J, Ekboir A, Mbabu D, Spielman D. Horna, and Benin, S. (2009). "From Best Practice to Best Fit: A Framework for Analyzing Agricultural Advisory Services Worldwide." Journal of Agricultural Extension and Education15 (4):341–55.

Christoplos I (2010). Mobilizing the Potential of Rural and Agricultural Extension. Rome: Food and Agriculture Organization of the United Nations (FAO) and the Global Forum for Rural Advisory Services (GFRAS)

Cecchini S, Scott C (2003). Can information and communications technology applications contribute to poverty reduction? Lessons from rural India, information technology for development, 10(2), 73-84.

Cieslikowsk DA, Halewood NJ, Kimura K, Zhen-Wei Qiang C (2009). Key trends in ICTdevelopment (World Bank Report). Retrieved August 7, 2010, from the Communication Initiative Network website: www.comminit.com/en/node/298770/307. Djankov, S., McLeish, C., Nenova, T., & Sheifer, A. (2001). Who owns the media? Journal of Law and Economics, 46(2).

Davis K, Heemskerk W (2012). Investment in Extension and Advisory Services as Part of Agricultural Innovation Systems. World Bank. In Agricultural Innovation Systems: An Investment Sourcebook 179-260). Washington, DC: World Bank

Dodds T (1999). Non-formal and adult basic education through open and distance learning in Africa. University of Namibia, Center for External Studies

Fafchamps M, Minten B (2011). Impact of SMS-based agricultural information on Indian farmers, The World Bank Economic Review, 1-32, Open University Press, Oxford. doi: 10.1093/wber/1hr056 [10]

Gakuru M, Winters K, Stepman F (2009). Inventory of Innovative Farmer Advisory Services using ICTs. The Forum for Agricultural Research in Africa (FARA).

Ilahiane H (2007). "Impacts of Information and Communication Technologies in Agriculture: Farmers and Mobile Phones in Morocco." Paper presented at the Annual Meetings of the American Anthropological Association, December 1, Washington, DC.

Lapar M, Holloway G, Ehui S (2003). Policy options promoting market participation among smallholder livestock producers: A case study from the Philippines. Food Policy, 28, 187–211

Katengeza SP, Mangisoni JH, Okello JJ (2010) The role of ICT-based market information services in spatial food market integration: the case of Malawi Agricultural Commodity Exchange, contributed paper presented at the Joint 3rd African Association of Agricultural Economists (AAAE) and the 48th Agricultural Economists Association of South Africa (AEASA)

Conference, Cape Town, South Africa, September 19-23, 2010. [20]

Kapange B (2012). ICTs and National Agricultural Research Systems .The case of Tanzania, unpublished paper, Ministry of Food Security and Cooperatives.

Kotable M, Helsen K (2001). Global marketing management. 2nd edition. New York: John Willey & Sons, Inc.

Labonne J, Chase RS (2009). "The Power of Information: The Impact of Mobile Phones on Farmers' Welfare in the Phillppines." World Bank Policy Research Working Paper No. 4996. Washington, DC: World Bank.

Lashgarara and Omidi (2011). ICT Capabilities in Improving Marketing of Agricultural Productions of Garmsar Township,Iran, Annals of Biological Research, 2011, 2 (6):356-363

Matambalya F, Wolf S (2001). The Role of ICT for the Performance of SMEs in East Africa, Empirical Evidence from Kenya and Tanzania, Discussion Papers on Development Policy No. 42, Center for Development Research, Bonn, December 2001, pp. 30.

Masuki KF, Tukahirwa J, Kamugisha R, Mowo J, Tanui J, Mogoi J, Adera EO (2012). Mobile phones in agricultural information delivery for rural development in Eastern Africa: Lessons from Western Uganda

Mittal S, Mehar M (2012), How Mobile Phones Contribute to Growth of Small Farmers? Evidence from India, Quarterly Journal of International Agriculture 51 No. 3: 227-244

Matungal P, Lyne, M, Ortman G (2001). Transaction costs and crop marketing in the communal areas of Impendle and Swayimana, KwaZulu Natal. Development Southern Africa, 2001; 18(3); 347-363

Munyua H, Adera E, Jensen M (2008), "Emerging ICTs and their potential in revitalizing small scale agriculture in Africa". A paper presented at World Conference on Agricultural Information and Information Technology

Mwakaje, A (2010). Information and Communication Technology for Rural Farmers Market Access in Tanzania, Journal of Information Technology Impact Vol. 10, No. 2, pp. 111-128, 2010

Mtega, W, Msungu, A (2013), Using Information and Communication Technologies for Enhancing the accessibility of agricultural information for Improved agricultural Production in Tanzania, The Electronic Journal of Information Systems in Developing Countries, Volume 56, 1, pages 1-14

Munyua H (2008). ICTs and small-scale agriculture in Africa: a scoping study. International Development Research Centre (IDRC) Policy brief, realizing the potential of ICT in Tanzania.

Asenso-Okyere K, Mekonnen DA (2012). The Importance of ICTs in the Provision of Information for Improving Agricultural Productivity and Rural Incomes in Africa, working paper, United Nations Development Programme

Sanga C, Kalungwizi VJ, Msuya CP (2013). Building an agricultural extension services system supported by

ICTs in Tanzania: Progress made, Challenges remain, International Journal of Education and Development using Information and Communication Technology (IJEDICT), 2013, Vol. 9, Issue 1, pp. 80-99

Stien J, Bruinsma W, Neuman F (2007). How ICT can make a difference in agricultural livelihoods: The Commonwealth Ministers Reference Book: International Institute of Communication and Development (IICD). Retrieved from http://www.iicd.org/files/ICT%20agricultural%20livelihoos.pdf

Shaffril H, Hassan M, Hassan A, D'Silva J (2009). "Agro-based Industry, Mobile Phone and Youth: A Recipe for Success." European Journal of Scientific Research 36(1):41–8.

Shetto MC (2008) Assessment of Agricultural Information Needs In African, Caribbean and Pacific (ACP) States Eastern Africa Country Study: Tanzania. Ministry of Agriculture, Food Security and Cooperatives on behalf of the Technical Centre for Agricultural and Rural Cooperation (CTA). http://icmpolicy.cta.int/filesstk/Tanzania_Final-report-081209.pdf

Tanzania Communication and Regulatory Authority (TCRA), Report, 2014

Tofik I (2014). Use of Mobile Phone in Camel Marketing, The case of Babille District of Fafau Zone, Somali Region, Ethiopia, Unpublished Master's Thesis

World Bank (2011). ICT in Agriculture, connecting small holder to knowledge, network and institutions

USAID Briefing Paper (2013). Using ICT to Enhance Marketing for Small Agricultural Producers. Last updated May 2013.

13

Determinants of Market outlet Choice for Major Vegetables Crop: Evidence from Smallholder Farmers' of Ambo and Toke-Kutaye Districts, West Shewa, Ethiopia

*Chala Hailu[1] and Chalchisa Fana[2]

[1,2]Department of Agribusiness and Value Chain Management, College of Agriculture and Veterinary Science, Ambo University, Ethiopia.

This study was initiated to investigate factors affecting market outlet choices by smallholder farmers' in Ambo and Toke-Kutaye districts. A total of 150 sample households were randomly selected for an interview using a semi-structured questionnaire. The Descriptive statistics and multinomial logit regression model were used for data analysis. Hence, 49.33% of sampled respondents choice direct sell to market while the remaining 31.33% and 19.33% of respondents choice wholesaler and retailer channel respectively. On the other hand, the multinomial logit regression analysis result showed that family size and access to market negatively affecting choice of retailer channel. Similarly, dummy model farmer, education level, and access to credit decrease the probability choice of retailer channel while it increases probability choice of wholesaler channel. Livestock in TLU and access to market decreases the probability choice of wholesaler channel. Finally, the study suggested that being model farmer, allocating more land for vegetables production, efficient use of family labor, access to market, and access to credit services would help to enhance smallholders capacity to produce vegetables that aligned to improve vegetables value chain in the study areas.

Key words: Market Outlet Choice, Multinomial logit, Smallholder Farmers', Study Areas, Vegetables.

INTRODUCTION

Marketing of vegetables crops has paramount opportunities and challenges for smallholder farmers. Since the majority of smallholder farmers have subsistence production, marketing is underdeveloped and inefficient (MoFED, 2010). The lack of adequate storage facilities constitute another constraint to both marketing and food security and, large quantities of agricultural commodities produced by farmers tend to rot away un-marketed, further the raising of productivity by smallholder farmers has been the inability of most them to get linked into the super market chains (Kamara et al., 2002).

Ethiopia has a variety of vegetable crops grown in different agro ecological zones produced through commercial as well as small farmers both as a source of income as well as food. Hence, on average more than 2, 3999,566 million of tons of vegetables and fruits are produced by public and private commercial firms (EIA, 2012). According to the CSA report during the year of 2014/15 of *mehere* (winter) season, the volume of vegetables crop produced and the area covered for production is 884,849.36 quintals and 6,779.23 hectares respectively.

Corresponding author: Chala Hailu, Department of Agribusiness and Value Chain Management, College of Agriculture and Veterinary Science, Ambo University, Ethiopia. **Email**: Caalaa2012@gmail.com
Co-author e-mail: hinsarmufana@gmail.com

Out of this volume, onion took the share of 1,645.03 quintals (0.24%) and 1,211.27 hectares and tomato shared 549,615.15 quintals (0.67%) and 6,779.23 hectare of land (CSA 2015).

Having this potential, in the study areas vegetables like Tomato and Onion are widely grown and marketed. Farmers produce in two seasons using irrigation water and rainfall. Vegetables production in the study areas was constrained by shortage of seeds/planting materials, diseases and insect pests, poor postharvest handling and poor linkage to market and market information (Bezabih et al. 2014). Perhaps vegetables like tomato and onion attract good price, but suffer from high price volatility. Particularly with producers and traders revealed that the existing market condition and production planning doesn't suit the nature of vegetable products where farmers reported extremely low prices particularly for onion and tomato. So that on stand selling is common by brokers and wholesalers. Therefore, this study tried to address households' decision to choose market outlet for vegetables marketing by answering the following specific objectives to:

✓ describe the socio-economic and institutional factors of vegetables producers and;

✓ Estimate determinants of smallholder farmers' choice of major vegetables (tomato and onion) market outlet.

METHODOLOGY

Description of the Study Areas

This study was carried out in Ambo and Toke Kutaye districts of West Shewa zone of Oromia National Regional State. *Ambo district* is situated at 8°56'30" - 8°59'30" N latitude and 37° 47'30" -37°55'15" E longitude in central Oromia, Ethiopia, 110 km west of Addis Ababa. The district has 34 rural kebeles of which 23 of them are vegetable producers, and Ambo is the capital of the district. The 2007 national census reported total populations for this district is 108,406, of whom 54,186 (49.98%) were men and 54,220 (50.01%) were women (CSA, 2007). On the other hand, *Toke Kuatye* is located between latitude of 08° 59' 01.1' N and longitude of 37° 46' 27.6' E. The district has 31 rural kebeles of which 28 of them are vegetables producer, and *Guder* is the capital town. The total human population of the district is 119,999, of which 59,798 (49.83%) were men and 60,201 (50.17%) were women; and 15,952 or 13.29% of its population were urban dwellers (CSA, 2007).

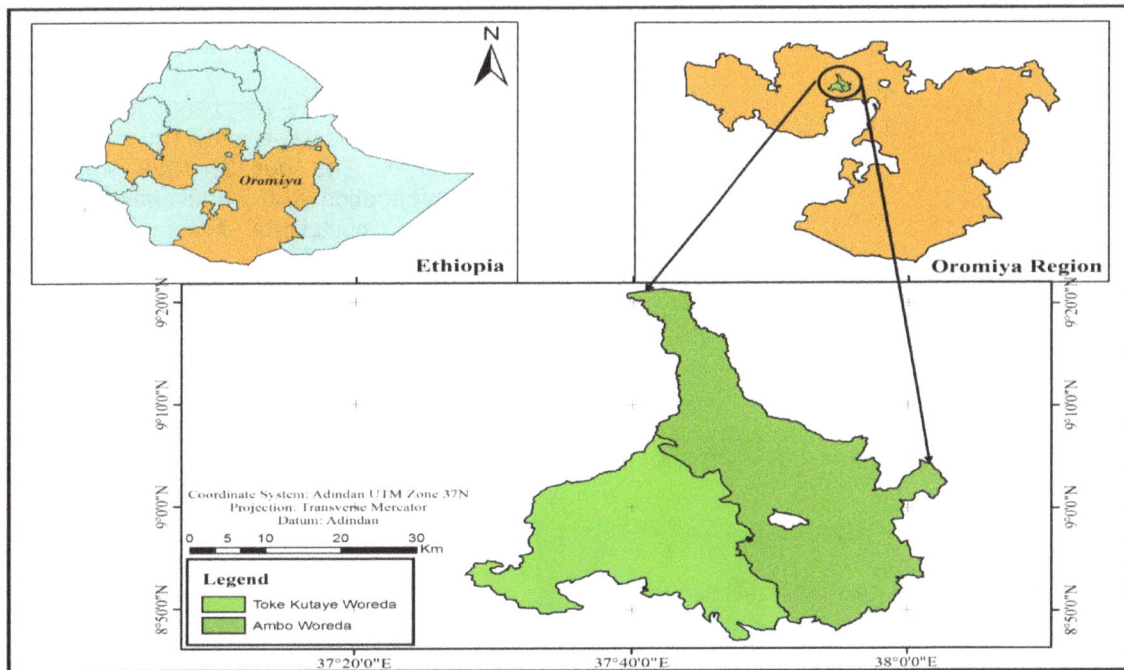

Figure 1. Map of the Study Areas (*Source: College of Agriculture, & Veterinary Science GIS laboratory*)

Major vegetables production and types irrigation systems in the study areas

Vegetables production is carried out in 28 and 23 kebeles of the Toke-Kutaye and Ambo district using irrigation water respectively. Among these kebeles, *Imala-Dawoo-Aajo*, *Birbirsa*, *Billo* and *Kiba* are ranked as the most widely vegetables producer. Hence, the modes of irrigation practices in the districts are traditional, motor pump,

modern irrigation scheme and water harvesting. For instance, there are 80 motor pumps and 20 modern irrigation schemes in Toke-Kutaye that are used by the farmers. The practice of traditional irrigation system is based on accessibility to water and farm field is irrigated using family labor. While farmers form a group of 5-6 and lay-out certain rules that function to use fairly the resource and settle conflict of interest. They purchase motor pump in group by raising their own fund and harvest water

Table 1. Types of Irrigation Practices by Districts

S/N	Types of Irrigation Practice	Total irrigable vegetable land in hectare by district		Total beneficiaries (Households)by district	
		Toke-Kutaye	Ambo	Toke-Kutaye	Ambo
1	Modern Irrigation System	1,495.5 (43.56%)	500.5 (13.38%)	3,461(53.23%)	2,685 (16.94%)
2	Traditional Irrigation System	1,618.25 (47.14%)	2747.4(73.46%)	2,457 (37.79)	10,734(67.71%)
3	Motor Pump Irrigation system	319 (9.30%)	492 (13.16%)	584(8.98%)	2434(15.35%)
Total		3,432.75	3,739.9	6,502	15,853

Source: *Ambo and Toke-Kutaye Districts Agricultural Office (DAO), 2016*

Moreover, the two districts have potential to produce different types of vegetables crop. Two season vegetables production using irrigation water and rain-fed is the predominant one. Tomato is dominantly produced in Toke-Kutaye district while cabbage is relatively the least. On the other hand, onion is largely produced in Ambo district. It is pertinent that, farmers pretended to produce following one another rather than market driven. Generally, the following are the major vegetables crop produced by year 2016/17 in Ambo and Toke-Kutaye districts.

Table 2. Major Vegetables Produced in the Study Areas

S/N	Major Vegetables crop	Total cultivated area (in ha)		District level per hectare productivity(in quintals)	
		Toke-Kutaye	Ambo	Toke-Kutaye	Ambo
1	Tomato	142,290	27,867	153	130
2	Onion	139,647	37,813.5	149	135
3	Irish Potato	100,980	160,643.9	153	158
4	Cabbage	44,928	32,432.4	108	108

Source: *Ambo and Toke-Kutaye District Agricultural Office (DAO), 2016*

Sampling Techniques and Procedures

For this study, three-stage sampling technique was employed. In the first case, sample districts were selected purposively based on the potential production of vegetables crops. Secondly, out of 23 and 28 kebeles of potential vegetable producers from Ambo and Toke Kutaye districts, 4 kebeles were selected based on their potential vegetables production namely *Kiba*, and *BillofromAmbo*, and *Imala DawoAajo* and *Birbirsa* from *Toke Kutaye*. Thirdly, 82 and 68 farm households (irrigation beneficiaries) were selected from Ambo and Toke-Kutaye district respectively based on Probability Proportional to Sample size using Yamane formula that resulted to 150 sample households (Yamane, 1967) i.e.

$$n = \frac{N}{1+N(e^2)}$$; Here the sampling error is 8% (0.08)

considering the budget, accuracy and time utilization for the research.

Table 3. Summary of Sample Kebeles by Respective Sample Households

District	Sampled Kebele	Total Irrigation Vegetable producers	Sample Size using PPS
Ambo	Billo	809	35
	Kiba	610	26
Toke-Kutaye	Imala Dawo Aajo	1585	68
	Birbirsa	496	21
Total		3,500	150

Methods of Data Collection and Data Sources

In this study, both primary and secondary data was used. To secure primary data, a semi-structured questionnaire was employed. The questionnaire was designed to capture information on household socio-economic characteristics, institutional factors, market information services variables and other major vegetables (Onion and Tomato) marketing related factors were included. Before conducting the survey data collection, pre-testing questionnaire was carried out. Thus, further revision was made in assuring the incorporation and deletion of the necessary and unnecessary factors. In addition, personal observation and focus group discussions were used to supplement information collected from respondents. Secondary data was also collected from District Agricultural Office (DAO) and other sources.

Methods of Data Analysis

Descriptive analysis

The descriptive statistics such as frequency and percentages was used to describe households' specific characteristics, wealth characteristics and institutional characteristics of vegetables producers.

Econometric Analysis

It is known that farmers' decision to choice one market outlet or another is categorized as a function of a set of incentives and capacity that allow the fulfillment of individuals demand. Given that we have formulated channel selection as a three-alternative choice (direct channel, retailer channel and wholesaler channel.). We have applied the multinomial logit model to estimate marketing channel choice with discrete dependent variable. According to rational choice theory, we assume individuals choice mutually exclusive alternative marketing channels to maximize utility and will choose the channel with maximum expected utility given their socio-economic and demographic characteristics and relevant resource constraints.

The producer's market channel choice can be conceptualized using a random utility model (RUM). RUM is particularly appropriate for modeling discrete choice decisions such as between market channels. It is an indirect utility function where an individual with specific characteristics associates an average utility level with each alternative market channel in a choice set. In our sample, smallholder farmers' did sell vegetables product using different channels. Producers are mapped into three mutually exclusive channels: direct channel, Retailer channel and Wholesaler channel.

Let decision-maker choose from a set of mutually exclusive alternatives, j = 1, 2, ..., J. The decision-maker obtains a certain level of utility U_{ij} from each alternative. The discrete choice model is based on the principle of that the decision-maker chooses the outcome that maximizes the utility. The smallholder farmers make a marginal benefit-marginal cost calculation based on the utility achieved by selling to a market channel or to another. We do not observe his/her utility, but observe some attributes of the alternatives as faced by the decision-maker. Hence, the utility is decomposed into deterministic (V_{ij}) and random (ε_{ij}) part:

$$U_{ij}=V_{ij}+\varepsilon_{ij} \forall ij N \tag{1}$$

A smallholder farmer selects market channel j=1 if

$$U_{ik}>U_k \tag{2}$$

Where U_{ik} denotes a random utility associated with the market channel j=k, and V_{ij} is an index function denoting the smallholder farmers' average utility associated with this alternative. The second term ε_{ij} denotes a random error which is specific to a producer's utility preference (McFadden, 1974). Now, in our implementation model, market channel choice is modeled as:

$$M_{ij}=\beta_j X_{ij}+\varepsilon_{ij} \tag{3}$$

Where M_{ij} is a vector of the marketing choices

(j = 1 for Direct channel; 2 for Retailer channel; and 3 for wholesaler channel of ith smallholder farmer, β_j is a vector of channel-specific parameters. ε_{ij} is the error term assumed to have a distribution with mean 0 and variance 1. X_{ij} is a vector of smallholder farmers' characteristics that together reflect the incentive, risks, capacity variables and other shifters influencing the producer's indirect utility, and hence his/her market channel decision.

Let Y be the unordered categorical dependent variable that takes on a value of zero or one for each of the J choices. The model for choice of supply channel can be given by:

$$\Pr(Y_i=j)=\frac{\exp(\beta_j'X_i)}{\sum_{j=0}^{J}\exp(\beta_j'X_i)} \text{ for j=1,2,3} \tag{4}$$

Where:

$\Pr(Y_i=j)$ is the probability of choosing either direct channel, retailer channel and wholesaler channel th direct channel as the reference supply channel strategy category,

J is the number of supply channel in the choice set,

j = 1 is direct channel, j= 2 is retailer channel, j = 3 is wholesaler channel.

X_i is a vector of explanatory factors conditioning the choice of the jth alternatives,

β is a vector of the estimated parameter.

The estimated equations provide a set of probabilities for the J + 1 choice restricted for a decision maker with characteristics. In order to remove an indeterminacy in the model, a convenient normalization that solves the problem is $\beta0 = 0$. Therefore, one can define the general form of the probability that individual i^{th} choose the alternative j^{th} in the following way:

$$\Pr(Y_i=j/X_i)=\frac{\exp(\beta'jX_i)}{1+\sum_{j=0}^{J}\exp(\beta'_jX_i)'} \text{ for all j >0} \tag{5}$$

The MNL coefficients are difficult to interpret and associating the β_j with the j th outcome is tempting and misleading. To interpret the effects of explanatory variables on the probabilities, marginal effects are usually used and derived as (Greene, 2003):

$$\delta_j=\frac{\partial P_j}{\partial X_j}=P_j\left[\beta_j-\sum_{j=1}^{j}P_j\beta_j\right]=p_j\left[\beta_j-\bar{\beta}\right] \tag{6}$$

The marginal effects measure the expected change in probability of a particular outcome being made with respect to a unit change in an explanatory variable (Greene, 2003).

Table 4. Summary of the Variables and Expected sign for Multinomial logit Model

Independent variables	Variables Code	Type of variables	Dependent variable = Choice of Market Outlet		
			Direct Channel	Retailer channel	Wholesaler channel
Age of the respondent	AGE	Continuous	+	+	-
Sex (1=male;0=female)	SEX	Dummy	-	-	+
Educational level in years of schooling	EDUCA	Continuous	-	-	+
Adult number of Family size	FMSIZE	Discrete	-	+	+
Access to market(1=yes;0=No)	ACMARKET	Dummy	+	+	+
Membership of cooperative	MEMCOP	Dummy	-	-	+
Owning Donkey for transportatlon(1=Yes;0;No)	OWDONKEY	Dummy	-	-	+
Distance from Main Public Road in time	DISROAD	Continuous	-	-	-
Number of livestock in TLU	LIVSTOCK	Continuous	+	+	-
Access to Credit Services (1=yes; otherwise '0')	ACREDIT	Dummy	-	+	+
Village level Status of the HH (1=model farmer; otherwise '0'	VLSHH	Dummy	-	-	+

RESULT AND DISCUSSION

Vegetables Market Outlet Choice in the Study Areas

Farmers choose different market outlet to supply their vegetables product. The identified vegetables supply channel is direct sell to market retailer and wholesaler.

Thus, 49.33% (n=74) of rural vegetable producer of the study areas sell their vegetables directly to the available local market where as the remaining 19.33% (n=29) and 31.33% (n= 47) were supplied their vegetables through the retailers and wholesaler channels respectively (Table 5).

Table 5. Percentages of Households by Vegetables Market Outlet

Vegetables market outlet	Districts				Total	
	Ambo		Toke-Kutaye			
	Freq.	Percent (%)	Freq.	Percent (%)	Freq.	Percent (%)
Directly sell to market	49	59.76	25	36.76	74	49.33
Retailers	10	12.19	19	27.94	29	19.33
Wholesaler	23	28.05	24	35.30	47	31.33
Total	**82**	**100**	**68**	**100**	**150**	**100**

Source: Own survey data (2016/17)

Socio- Economic Characteristics of the Households

As depicted in Table 6, the large number of respondents fall in age category between 35-45 years in which 46.34% (n=38) of them were from Ambo district, 35.29% (n=24) from Toke-Kutaye and 41.33% (n=62) for the whole sample. We have similar figure and percentage result for the lower and upper age category for whole sample. However, we found high number of respondents' age between 25-35 years in Ambo district and lower age figure in Toke-Kutaye district. In addition, the upper age category above 55 years was higher in Toke-kutaye district as compared to Ambo district. From the total of 150 sample households, male households were 84 %(n=126) and females were 16 %(n=24). We observed equal number of males in both study areas. The female households were larger in Ambo relative to Toke-Kutaye district. In addition, the majority of the households were literate while smaller numbers of households were illiterate. Generally, in both districts there was insignificant difference between number

of literate and illiterate. On the other hand, on the basis of adopting full agricultural package and join for commercialization households were grouped into model farmer and non-model farmer and found 36.67 %(n=55) model farmer and 63.33 %(n=95) non-model farmer. We have large number of model and non-model farmers from Ambo district relative to Toke-kutaye district. Likewise, family size of the households also assessed and the large number of households have family size between 5 and 8 in number where as small number of households have more than 8 family members. Finally, households asset endowment such as oxen and land holding were described accordingly, and 66.67 %(n=100) households have more than 1 *timad* of oxen and the remaining 32 % (n=48).and 1.33 %(n=2) households have 1 *timad* of oxen and no oxen respectively. On the other hand, 64.66 %(n=97) of households have more than 2 hectare of land where as 34.67 % (n=52) have between 1 and 2 hectare hence, 0.67% (n=1) have below 1 hectare of land.

Table 6. Socio-economic Characteristics of the Households

Socio-Economic Characteristics	Districts				Total (N=150)	
	Ambo (n=82)		Toke-Kutaye (n=68)			
	Freq	Percent	Freq.	Percent	Freq.	percent
Age Category						
25-35years	16	19.51	9	13.23	25	16.67
35-45Years	38	46.34	24	35.29	62	41.33
45-55years	18	21.95	19	27.94	37	24.67
≥ 55 years	10	12.19	16	23.53	26	17.33
Sex Category						
Male	63	76.83	63	92.65	126	84
Female	19	23.17	5	7.35	24	16
Educational level						
Basic education - 5 years	72	87.80	62	91.18	134	89.33
Illiterate	10	12.19	6	8.82	16	10.67
Household farm status						
Model Farmer	31	37.80	24	35.29	55	36.67
Non-Model Farmer	51	62.20	44	64.71	95	63.33
Family size category						
≤ 4 in number	27	32.93	12	17.65	39	26
5-8 in number	44	53.66	41	60.29	85	56.67
> 8 in number	11	13.41	15	22.06	26	17.33
Number of Oxen in *timad*						
No oxen	1	1.22	1	1.47	2	1.33
1 *timad* of Oxen	24	29.27	24	35.29	48	32
More than 1 *timad* of Oxen	57	69.51	43	63.23	100	66.67
Total Land Holding						
Less than 1 hectare	0	0	1	1.47	1	0.67
1- 2 hectare	39	47.56	13	19.12	52	34.67
More than 2 hectare	43	52.44	54	79.41	97	64.66

Source: Own computation data (2016/17).

Households access to institutional and infrastructure services

Institutional and infrastructure factors play significant role for vegetables producers. From the total respondents 73.33% (n=110) have access to irrigation agricultural extension services where as 26.67% (n=40) do not access to irrigation agricultural extension services. On the other hand, 94% (n= 141) and 6% (n= 9) have access to credit services. Finally, 76.67%(n=115), 54% (n=81) and 64% (n=96) were access to market place, membership of agricultural cooperative and access to public road showing that fortunate condition to produce vegetables. However, information from focus group discussion, justified that the lack of extension service on disease management damaging the quality vegetables (Table 7).

Table 7. Households Access to Institutional and Infrastructure Factors

Institutional and infrastructure	Yes		No		Total	
	Freq.	Percent	Freq.	Percent	Freq.	percent
Access to irrigation agricultural extension services (AIAEXT)	110	73.33	40	26.67	150	100
Access to credit services for vegetables production (ACREDIT)	141	94	9	6	150	100
Access to market place (ACMARKET)	115	76.67	35	23.33	150	100
Access to main public road (ACROAD)	96	64	54	36	150	100
Membership of agricultural cooperative (MEMCOP)	81	54	69	46	150	100

Source: Own survey data (2016/17)

Determinants of Market Outlet Choice for Major Vegetables

In the next econometric analysis, multinomial logit model (MNL) was employed to estimate the determinants of vegetable market outlet choice by households in setting three options of vegetable market outlet in the model. During the procedural estimation direct sell channel was tailored to the reference category based on highest rate of respondents' choice. We incorporate 12 explanatory variables in the model and interpreted at the marginal effects. The goodness fit of the model is 0.013 which is significant at 5%. Likewise, heteroskedasticity and multicollinearity test was conducted for the variables used in the model and found no serious problem i.e. less than 10%.

Model Farmer (MOFARM): A dummy model farmer variable significantly determining the vegetables channel choice. Being a model farmer decreases the probability choice of retailer channel by 17.9% and increases the probability choice of wholesaler channel by 11.2%. This indicates that model farmer produces the bulk of major vegetables (onion and tomato) using given resources and farming experience. This may help to trap better market price and that induces to supply bulk vegetables product to wholesalers.

Educational status (EDUCA): The dummy educational status of the household is important variable affecting the vegetables market channel choice. Hence, literacy decreases the probability to choose the retailer channel for vegetables marketing and increases to choose wholesaler market channel. It is significant and affects retailer and wholesaler market channel choice at 10% and 1% probability level respectively. This may be due to literate households are more aware of market channel and able to get market information for their produce and helps to choose the best market channel that expected to give better price for their produce. The result in this finding is consistent to Abrahm, (2013), in his analysis of vegetables market outlet choice in Habro and Kombolcha districts.

Total Family Size (FMSIZE): It is positively affecting the probability choice of the retailer market channel and consistent with the hypothesis set. An increase in number of family member increases the probability choice of retailer market channel by 2.8% and significant at 10% probability level. This indicates that the more family size helps to supply vegetables to different retailer shops, restaurants and kiosks in different units which affects to operate vegetables production.

Total Number of Livestock Owned (TLIVSTOK): Livestock are important in contributing household income.

The sign obtained for this variable also consistent with the hypothesis thus an increase in number of total livestock unit in TLU reduces the probability choice of wholesaler market channel by 2.4% and significant at 5% probability level. This may indicate that households who have more livestock allocate more of their land for grazing area and fodder production using irrigation water which tends to reduce land used for vegetable production. Information from group discussion confirmed that the shortage of open grazing land in the districts especially during winter season allows cattle to graze over wet part of their land. Hence, it reduces vegetables production at bulk that purchased by wholesalers.

Access to Credit Services (ACREDIT): The dummy access to credit services affect negatively the probability choice of retailer market channel. Access to credit services decreases the probability choice of retailer channel by 13.8% and significant at 1% probability level. The sign of the finding is opposite to the hypothesis indicating that credit is advantageous to produce the vegetables at bulk that rarely marketed to retailers. This shows that the more households acquire credit services, the more they increase scale of vegetables production. Because credit facilitate fortune condition to acquire inputs such as motor pump, water can and other inputs that leads to produce more vegetables (Onion and tomato) which attracts wholesaler. This result is consistent with Mebrat (2014) in which she obtained credit has positive effect in vegetables production in rift valley of Ethiopia.

Access to Local Market Area (ACMARKET): Due to their minimum shelf life as well as risk in product loss, vegetables production should be near to public road and market areas. Thus, dummy access to market area increases the probability choice of retailer market channel by 11.3% and significant at 5% probability level. This may indicate that the more households access to market area, the more diversify their vegetables production on their limit land and supply to retailers. This finding is contrary to Bezabih et al, 2015, who found negative sign for retailer channel in their study of potato value chain. On the other hand, access to local market area reduces the probability choice of wholesaler market channel by 21.3% and significant at 5% probability level. This may indicate that there may be land and water shortages to produce vegetables at large and given that all water users busy their schedule to water their vegetables farm field. The significant result obtained for wholesaler is opposite to (Shilpi and Umali, 2007; Sirak and Bahta, 2007) who obtain positive sign to the channel and similar to Tewodros (2014) in his market outlet analysis of chickpea in southern Ethiopia found negative sign for wholesale market participation.

Table 8. Marginal Effect Estimates at Mean from Multinomial Logit Model

Variables	Retailer Channel			Wholesaler Channel		
	dy/dx	Std.Err.	P>Z	dy/dx	Std.Err	P>Z
MOFARM[d]	-0.179	0.076	0.019**	0.250	0.112	0.026**
AGE	-0.003	0.002	0.156	-0.009	0.003	0.782
SEX[d]	-0.039	0.070	0.577	0.008	0.093	0.924
EDUCA[d]	-0.259	0.138	0.060*	0.242	0.069	0.001***
FMSIZE	0.028	0.010	0.070*	-0.024	0.018	0.192
TLIVSTOK	0.006	0.008	0.298	-0.024	0.119	0.042**
ACREDIT[d]	-0.138	0.044	0.002***	0.031	0.092	0.736
ACMARKET[d]	0.111	0.047	0.019**	-0.213	0.099	0.033**
DISROAD	-0.048	0.046	0.300	0.002	0.021	0.918
MEMCOP[d]	-0.055	0.050	0.280	0.001	0.087	0.985
USDONKEY[d]	0.077	0.052	0.138	0.028	0.091	0.756

The table header "Market Outlet Choice (Direct sale to market = reference category)" spans both channels.

Source: own calculation from STATA ver. 13 (*) Significance levels of 10%, (**) Significance levels of 5%, (***) and Significance levels of 1%, ([d]) dy/dx is for discrete change of dummy variable from 0 to 1

CONCLUSIONS

Identifying factors affecting vegetables farmers' market outlet choice is important for the development of vegetables value chain. The volume of vegetables production plus farmers' productive assets determines what channel to choose. Hence, the less the volume of vegetables produced the more they choose retailer and vice-versa. In general, farmers' status, wealth and institutional factors contribute a lot for marketing of vegetables. And most farmers directly sell to market and relative numbers of them choose wholesaler channel and followed by retailer channel. Therefore, the study can contribute literature that helps to develop vegetables value chain particular to the study areas.

ACKNOWLEDGMENTS

First and foremost, we would like to express our gratitude to almighty God for enabling me to complete this research work successfully Next, we have strong gratitude to Ambo University who gave me an opportunity to fund the research and make it smoothen to complete the work. Finally, we have also recognition for the deliberation of staff members of Agribusiness and Value Chain Management department for all achievements.

REFERENCES

Abraham Tegegn(2013). Value Chain Analysis of Vegetables: The Case of Habro and KombolchaWoredas in Oromia Region, Ethiopia. Msc Thesis Submitted to Haramaya University unpublished.

AyelechTadesse(2011). Market Chain Analysis of Fruits for Gommaworeda, Jimma zone, Oromia National Regional State. An M.Sc Thesis Presented to School of Graduate Studies, Haramaya University.pp110.

Baltenweck, I. L., Njoroge, R., Patil, M. I. andKariuki E.(2006). Smallholder Dairy Farmer Access toAlternative Milk Market Channels in Gujarat. Contributed paper IAAE Conference, Brisbane, Australia.

Bezabih,. E., Amsalu A,, Tesfaye,B., and Milkesa, T.(2014). Scoping Study on Vegetables Seed Systems and Policy in Ethiopia. Final Report, Addis Ababa, January 2014.

Bezabih,E., Mengistu K., Jeffreyson K. M. and Jemal,Y. (2015). Factors Affecting Market Outlet Choice of Potato Producers in Eastern Hararghe Zone, Ethiopia. *J. of Economics and Sustainable Development*. Vol 5, No.15.

Bongiwe, G. and B. Masuku (2012). Factors affecting the choice of marketing channel by vegetable farmers in Swaziland. Canadian Center of Science and Education. J. *Sustainable Agriculture Research*, 2 (1):123.

CSA (Central Statistical Authority) (2007). Summary and Statistical Report of 2007 Population and Housing Census. Federal Democratic Republic of Ethiopia Population and Census Commission.

CSA (Central Statistical Agency) (2015). Large and Medium Scale Commercial Farms Sample Survey. Result at Country and Regional Level Volume VIII. Statistical Report on Area Production of Crops, and Farm Management Practices, Statistical Bulletin, August, 2015, Addis Ababa, Ethiopia.

DAO (District Agricultural Office)(2016). District Irrigation Office Annual Report Document.

EIA (Ethiopian Investment Agency)(2012). Investment Opportunity Profile for the Production of Fruits and Vegetables in Ethiopia.

Girma, M. and Abebaw, D. (2012). Patterns and Determinants of Livestock Farmers' Choice of Marketing Channels: Micro-level Evidence. EEA/EEPRI working paper, January, 2012 Addis Ababa.

Greene, W.H.(2003). Econometric Analysis, 4th ed., Prentice Hall, pp.640 – 642.

Kamara A.B., B. Van Koppen, and L. Magingxa (2002), "Economic Viability of Smallscale Irrigation Systems in the Context of State Withdrawal: the ArabieSchemein the Northern Pronvince of South Africa". *J. Physics and Chemistry of the Earth*,27: 815-823.

Kuma, B., Derek, B., Kindie, G. and Belay, K. (2013). Factors Affecting Milk Market Outlet Choices in Wolaita Zone, Ethiopia. *African J.of Agricultural Research, 8 (21), 2493-2501.*

McFadden, D. (1974). The Measurement of Urban Travel Demand. *J. of Public Economics,* 3: 303-328.

MebratTola(2014). Tomato Value Chain Analysis in the Central Rift Valley: The Case of DugdaWereda, East Shoa Zone, Oromia National Regional State, Ethiopia. A thesis submitted to Haramaya University.

MoFED (Ministry of Finance and Economic Development), 2010. Growth and Transformation Plan (GTP) 2010/11-2014/15 Draft. September 2010, Addis Ababa.

Shilpi, F. and Umali-D, D.(2007). Where to sell? Market Facilities and Agricultural Marketing. Policy Research Working Paper series 4455, The World Bank.

Sirak, T. B. and Bauer, S. (2007). Analysis of the Determinants of Market Participation within the South African Small-Scale Livestock Sector. Trope tag paper, October 9-11, 2007.

Tewodros Tefera(2014). Analysis of Value Chain and Determinants of Market Options Choice in Selected Districts of Southern Ethiopia. Journal of Agricultural Science; Vol.6, No. 10; 2014, ISSN 1916-9752 E-ISSN 1916-9760, published by Canadian Center of Science and Education pp. 37.

World Bank(2011). ICT in Agriculture: Connecting Smallholders to Knowledge, Networks, and Institutions, Washington D.C.: World Bank.

Yamane, T. (1967). Statistics: An Introductory Analysis, 2nd Ed., New York: Harper and Row.

Zivenge, E. and Karavina, C. (2012). "Analysis of Factors Influencing Market Channel Access by Communal Horticulture Farmers in Chinamora District, Zimbabwe." *J. of Development and Agricultural Economics.* Vol. 4(6), pp. 147-150, 26 March, 2012.

14

Contracting decision and performance of Mexican coffee traders: The role of market institutions

Benigno Rodríguez-Padrón[i], André Ricardo Cortés Jarrín[2] and Kees Burger[3]

[1]Chapingo Autonomous University,Chapingo, State of Mexico.
[2,3]Development Economics Group, Wageningen School of Social Sciences, Wageningen University, Hollandseweg, The Netherlands.

We identified and explained the contractual choices of Mexican coffee traders in selling their product and analyzed the traders´ performance. The data were obtained from personal interviews with 53 intermediaries in four coffee producing regions of the states of Oaxaca and Veracruz, Mexico. Marketing margins were used as an indicator of traders' performance. The results indicate that being a roaster, having a wet processing plant and selling cherry coffee negatively affects the use of contracts whereas being vertically integrated has a positive effect. The results also suggest that being registered in the National Coffee System (which only a minority of the interviewed traders were) increases the margin for the trader. Selling cherry coffee, participating in a competitive environment and having a contract decreases these margins (at 5% significance) and may thus enhance the performance of the supply chain and benefit the producers.

Keywords: Contractual arrangements, intermediaries, trade, coffee, Mexico.

INTRODUCTION

Middlemen play an important role in a commodity chain. They perform many of the activities required to bring goods from producer to consumer, and their productivity has a strong impact on the performance of the chain as a whole and the welfare of the agents involved (Sexton and Lavoie, 2001). Differences in the behavior of middlemen can be explained by economic and cultural factors. The way intermediaries interact can give information about the environment in which they make their transactions and vice versa (Fafchamps, 2004; Fafchamps and Minten, 1999).

Relationships and the environments in which intermediaries act are not static; they evolve over time. Agricultural commodity markets have undergone significant changes over the past twenty years, changes that at the same time affect the behavior of agents within a commodity chain. This is the case for the coffee chain, too. The major change occurred at the end of the nineteen eighties and

early nineties, when the coordinated marketing system under the International Coffee Agreement was abandoned. In many countries, including Mexico, this led to a withdrawal of the government from the sector, creating scope for private parties and their organizations. Since then, the degree of state interference has fluctuated; private sector organizations or public-private initiatives have come and gone.

*Corresponding Author: Benigno Rodríguez-Padrón, Universidad Autonoma Chapingo, kilometre 38.5 road, Mexicoto Texcoco. Chapingo, State of Mexico. E-mail: beroopadron_67@hotmail.com. Supported by the Alban Programme, the European Union Programme of High Level Scholarship for Latin America, scholarship number E06D100933MX; and by the Ministry of Science and Education of Mexico (CONACYT).

In this paper, we look at a particular aspect of the institutional arrangements, namely the use of sales contracts by the agents who buy coffee from the farmers or from other buyers.

There is a variety of channels in the Mexican coffee supply chain; within these there is a variety of steps through which coffee transits from producer to consumer (see Annexes 1 and 2). The simplest one is the product sold directly from the producer to the consumer. In this case, the product is sold at the regional, local or national market. Yet, the most common channel is the one starting from the producers who sell to the local collectors; then there are the regional collectors, and after that there are the state buyers; after this, the coffee is transported to the border to be sent to other countries by exporters. From then on, the product is taken over by processors (roasters) of consumer countries. This description implies that there are five steps in the supply chain, in this study we interviewed intermediaries placed on the first four steps.

Studies show that the relationships with middlemen can be crucial for the performance of the commodity chain (Gabre-Madhin, 1999;Fafchamps*et al.*, 2005).Knowledge of the Mexican coffee intermediaries' behavior, contributes to a better understanding of the coffee market, and gives information to improve Mexican coffee policies. Therefore, in this paper we investigate the coffee traders´ behavior and the performance of the Mexican coffee agents. We focus on two coffee-producing states, Oaxaca and Veracruz.

More in particular, the objectives of this work are: to identify and explain the contractual sales arrangements of traders in the Mexican coffee chain and to assess the performance of coffee traders in the states of Oaxaca and Veracruz, Mexico. In order to reach these objectives, the following research questions need to be answered: what are the socio-economic characteristics of coffee traders in the states of Oaxaca and Veracruz, Mexico? What are the main factors that affect whether traders have a contract with their buyers? And how do contracts influence the performance of coffee traders in Oaxaca and Veracruz, Mexico?

The data used in our analysis were obtained by conducting face-to-face interviews with 53 intermediaries in four coffee-producing regions of the states of Oaxaca and Veracruz. Additionally, we held meetings with local authorities and people involved in the Mexican National Coffee System. Figure 1 shows that coffee prices have fluctuated enormously over the past decades, and that the margins between fob (free on board) prices and producer prices were also quite variable.

The margins range from 2.30% of the selling price when buying and selling cherry coffee to 63.48% when buying cherry and selling roasted coffee. We use this margin as a measure of the performance of the intermediary. Contractual arrangements with their buyers may help intermediaries invest in specialized equipment or otherwise reduce costs of trading and/or processing. But many transactions are not based on contracts. The transaction costs of arranging a contract may exceed the benefits of having one in cases where reality is difficult to capture in regulations, or where contact is already intensive. We elaborate on these reasons in the next section.

The empirical work indicates that being a roaster, having wet processing facilities and selling cherry coffee negatively affect traders' use of contracts, whereas having dry processing facilities has a positive effect on their use of contracts. The findings also suggest that selling cherry coffee, participating in a competitive environment and having contracts positively influence intermediaries' performance, while being registered in the National Coffee System have a positive effect on the gross margin. The remainder of this paper is organized as follows: in Section 2, we present the theoretical approach; in Section 3, we present the data and methods used to analyze the information; in Section 4, we give the model specification and likely determinants of contract choices; in Section 5, we specify the model of traders´ performance and show the empirical findings; and in Section 6, we present the conclusion.

THEORETICAL APPROACH

Why are institutions so important for understanding contractual choices and the performance of traders in chains? According to North (1989), every process in which an exchange is performed has some costs involved. In the case of the Mexican coffee sector, one can think of the search for market information, finding farmers who sell coffee, inspecting the quality and transporting the product and arranging contracts between the agents who perform the mentioned activities (Jabbar*et al.*, 2008). Contracts arise as facilitators of these interactions; they provide the possibility of doing business at low costs and thus influence the efficiency along the commodity chain. If transaction costs are low, economic actors will favor spot markets, but if transaction costs are high, they will favor contracting or vertical integration to lower these costs (Ruben *et al.*, 2007a). Additionally, Janvry *et al.*(1991) have argued that the existence of transaction costs might explain why households are not entering a certain agricultural market. Transaction Cost Economics (TCE) helps to understand how agents decide to run their business given the environment in which they execute their activities. Some traders may choose to buy and sell coffee without adding much value to the product they market, as costs and uncertainty are high. Others will choose to vertically integrate and sell more downstream in order to tackle these costs. Some agents turn to contractual agreements in order to reduce these costs (Williamson, 1979).Three principle attributes of transactions have been identified: the frequency with which transactions recur, the specificity of the assets necessary to come to a transaction, and the degree and type of uncertainty of transactions. Normally local intermediaries make contractual arrangements and the roasters and exporters go for vertical integration. The TCE theory predicts that under high asset specificity and high uncertainty, the firms will embrace a highly integrated channel in all cases (Shervani et al.,2007); thus more contracts should be found for agents with specific assets, uncertainty and frequency of transactions.

In terms of contracts, TCE portraits contracts as methods to constrain ex-post behavior, given the fact that there may be opportunistic behavior, information asymmetries, differences in bargaining power, and possible hold-up situations from asset-specific investments among agents (Williamson, 1979). TCE focuses as well on the determinants and the duration of contracts. It also distinguishes between a contractual and a non-contractual exchange, which has implications in terms of the formality of relationships (Masten and Saussier, 2000). Processing facilities, such as wet and dried plants, processing and roasting machines are specific assets for coffee, but not always for a specific buyer which implies that traders who have these facilities will prefer to have a contract with their buyer; in this way, they reduce uncertainty and opportunistic behavior. In fact a roasting facility provides an opportunity to serve many more customers and so leads away from being dependent on single buyers. Thus, having a multitude of customers compensates for a contract.

Traders face an uncertainty due to price volatility in the coffee market. Considering this possibility, intermediaries may want to have a contract with their buyer in which they can negotiate schemes to share the risks of a volatile market before the harvest season (and therefore the buying season) starts. This kind of agreements, in which uncertainty is reduced, may give them a higher utility. However, not all coffee traders face such uncertainty. There are traders who do not rely on coffee to make their living because they have another business parallel to marketing coffee. If traders can diversify their sources of income, they have a reason for not having a contract. Moreover, traders who have different lucrative activities usually do not have coffee asset-specific investments, hence another reason for not having a contract.

Two parts are involved in the transaction costs. One part is the ex-ante cost, which includes searching for potential exchange agents (consumers or wholesalers who offer the best price), the screening of potential agents, and bargaining (Keyet al.,2000). The second part consists of the ex-post costs that take into account the transfer of property rights and the monitoring of compliance of any transaction (Rubenet al.,,2007b). When markets are far from representing the ideal situation portrayed in economic textbooks, variables like trust, reputation and informal rules gain in importance (Fafchamps, 2002). As Gabre-Madhin (1999) asked, "in the absence of formalized market institutions that deter dishonest behavior, such as credit bureaus, trade inspection services, and commercial tribunals, what institutions arise that promote trade among unknown parties?" The answer was that the brokers help in doing so. These social relationships then have an effect on contractual decisions and on trader performance. If there is trust between agents, formal agreements may not be needed (Fafchamps and Minten, 1999). Furthermore, these relationships may serve as substitutes to absent market institutions and may help traders to reduce costs and perform efficiently.

The exertion of market power by some firms in the coffee sector may also affect the behavior of other intermediaries. There is some concern about the level of competition within the Mexican supply chain, given that agricultural markets can show evidence of low levels of concentration of buyers or sellers (Sextonand Lavoie, 2001). Competition is thus an important element in the analysis. Being in a competitive environment is a factor that increases the probability of contractual agreements between agents as they have to secure the provision of the product and the market in which that product will be sold (Fischeret al.,2009). Low competition may lead to higher marketing margins, which decreases the efficiency in the supply chain (Mose, 2007; Schroeter andAzzam, 1991).

An environment in which transaction costs are high may lead to higher margins, and thus, to a less efficient trader and supply chain. Poor physical infrastructure, high costs of processing, poor institutions and high costs of information gathering and monitoring contracts, are some of the costs associated with an inefficient set-up and higher margins (Jabbar et al., 2008; Winter-Nelson and Temu, 2002). On the other hand, in a competitive environment, market power and hence marketing margins are reduced, creating a more efficient industry (Porter, 1998; Schroeter and Azzam, 1991). In this sense, a liberalization process like the one experienced by the Mexican coffee sector should end up in a competitive structure in which marketing margins of traders are low.

The Mexican coffee sector went from being ruled by a state-led marketing board to a free market set-up. This meant that part of the role played by the government in the regulated era had to be taken up by private agents. This role included not only buying, processing and marketing the coffee from growers but also providing them with financial and technical assistance. Those changes meant that new arrangements between these agents also had to be established. They may do so in order to reduce the transaction costs and the risk in the coffee market. One of these arrangements is that traders, especially exporters, may be vertically integrated in order to reduce transaction costs (Mehta and Chavas, 2008;Williamson, 2000; Winter-Nelson and Temu, 2002).

Until the end of the nineteen eighties, the world market for coffee was controlled by a quota system resulting from the economic clauses of the International Coffee

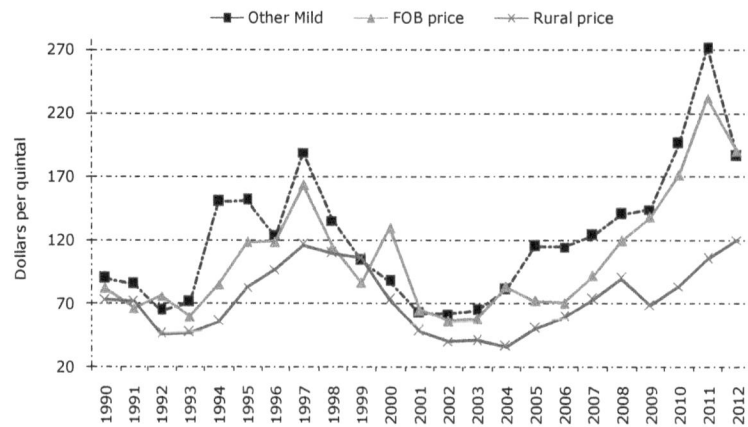

Figure 1. Prices along the coffee supply chain (own elaboration with data from ICO and SIAP).

Agreements (ICAs), signed by the main producing and consuming countries. These ICAs were first put into action in the early nineteen sixties. The agreements stated that producing countries had the responsibility to control their coffee exports in order to affect world prices (Gilbert, 1987). The best way to do this was by means of direct government intervention in the producing, processing and marketing of coffee. This is why, under this controlled regime, most of the producing countries had marketing boards controlled by the state. These public institutions had the monopoly of the coffee trade and some of them were also in charge of technical assistance and financial support to coffee growers (Akiyama, 2001).

In the Mexican case, the government created the Mexican Coffee Institute (Instituto Mexicanodel Cafe, INMECAFE) to control and promote coffee production and sales. The Institute sponsored this crop as a remunerative alternative for peasants via its different support programs(Santoyoet al.,1994). Given that coffee-producing areas were growing and spreading, the Institute had to make sure these regions remained under the control of the government. Hence, officials were sent and infrastructure was built in several producing regions (Pérez et al., 2001). While the INMECAFE and ICAs were working fully, there was no need for local or regional partner organizations, since every actor in the coffee supply chain relied on the effectiveness of the institutional environment until 1989.

Yet, at the end of the nineteen eighties, the ICAs' economic clauses broke down. This meant that the quota system was out, replaced by the free market. At the same time, liberalization policies were spreading throughout Central and Latin American countries, and Mexico was no exception. New coffee policies focused on reducing the size of the government and its expenditures, prioritizing regulation and macroeconomic stability, and abandoning interventions in commodity markets. For the Mexican coffee sector this meant INMECAFE had to

disappear (Snyder, 1999). The Institute was then dismantled between 1989 and 1992, creating a void in the coffee chain; a void that was to be filled by private agencies. In the absence of the state's marketing boards and direct involvement in the coffee market, new local, regional and national traders appeared.

The local small-scale traders still exist because large-scale firms lack the information the local and regional collectors have about the geographical area where they make their purchases. Small-scale traders know where the growers live and where the best coffee is sold. This means that new contractual arrangements were also created between large- and small-scale traders, depicting an agency relationship. In the first section of our analysis, we will investigate factors that may affect this contractual choice, elements relating to asset-specific investments, trade-credit relationships, and the type of coffee that is marketed by contracting partners.

A prime reason for wanting to secure a trade relationship is having made investments that are specific to these relationships. Traders who invest in machinery or in assets that can only be used in coffee-related activities may want to have contracts with their buyers in order to reduce the risk of being locked-in. In this sense, having a long-term contract may give them higher benefits, since they will not face ex-post opportunistic behavior from their buyers, who are aware of the traders' sunk costs due to asset-specific investments (Williamson, 1979).

However, traders who face recurrent transactions with their buyers, for example those who sell coffee every day, and who have not made any specific investments besides this, may not need a contractual agreement with their buyers. Repeated interactions between agents can lead to trust-like relationships that do not rely on formal agreements to secure transactions (Fafchamps and Minten, 1999).

Traders may also face asymmetries in information. Some large traders are unaware of the location of coffee growers or local collectors. With this in mind, small- or

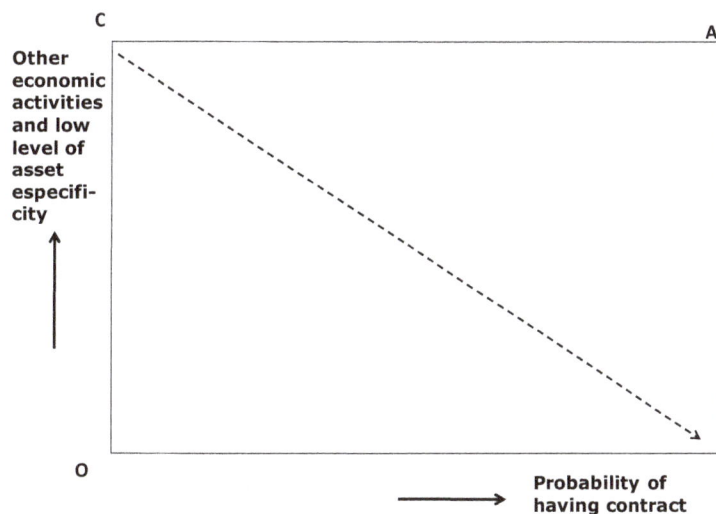

Figure 2.Relationship between other economic activities and contract.

medium-scale intermediaries who know where to buy coffee can signal large-scale buyers that they have this knowledge and may want to enter into a contract with them. In order to protect this information, intermediaries may want to negotiate better prices. If the intermediaries who know where to buy coffee, provide this information to a buyer without a contract, there is no guarantee that the buyer will buy the coffee from them. A contract may also reduce search costs involved in transactions and in this sense both parties could benefit utility from an agreement. Traders face an uncertainty that may be due to price volatility in the coffee market. Considering this possibility, intermediaries may want to have a contract with their buyer in which they can negotiate schemes to share the risks of a volatile market before the harvest season starts. This kind of agreements, in which uncertainty is reduced, may give them a higher utility. However, not all coffee traders face such uncertainty. There are traders who do not rely on coffee to make their living because they have another business parallel to marketing coffee. If traders can diversify their sources of income, they have a reason for not having a contract. Moreover, traders who have different lucrative activities usually do not have coffee asset-specific investments, hence another reason for not having a contract (see Figure 2).

Market institutions are a key factor when studying performance. The rules of the game can shape the way in which traders decide to run their business. In this analysis, we will use market institutions as a tool to understand traders' performance. These institutions can determine the way in which traders purchase and sell their coffee, how they pay and receive payment, and also the way in which they interact with other traders (Gabre-Madhin, 2001). The assets and the socio-economic characteristics of each trader, however, can also determine their trading practices. According to Jabbar et al. (2008), actors who face the same market institutions can have different trading practices, denoting that there

are trader-specific variables such as asset specificity, size and experience affecting trading behavior, and thus, their performance.

We have argued that new agents entered the sector as a consequence of liberalization. These changes, as we have explained, should furthermore impact traders' performance. From the fieldwork done for this study we observed that, after the liberalization, erstwhile exporters decided to become more involved in upstream steps of the supply chain; some of them acquired infrastructure that was left by the extinct INMECAFE. However, in the past twenty years, there has been little new investment in processing plants and infrastructure related to the coffee sector. This poses a constraint to the coffee chain, since processing plants are operating with old machinery. The lack of machinery to process coffee and roads in some regions also shapes the behavior of traders, since they have to take this condition into account in their trading practices.

We use marketing margins as an indicator of traders' performance as in Jabbar et al.(2008) and Mose (2007) since they can give an indication of how traders perform given the minimization of their variable costs. These margins are measured as the difference between the purchase and the selling price of a quintal of parchment coffee.[1] It is important to note that in our fieldwork we got information from traders who trade other types of coffee besides parchment coffee. In this logic, we had to make an adjustment in order to make the quantities and prices of coffee traded by each trader comparable. For this reason we transformed prices from other types of coffee into price of parchment coffee. This was done by applying a weight conversion; after this process, we obtained all prices in terms of quintals of parchment coffee purchased

---

[1] A quintal is 245 kg of cherry coffee, 57.5 kg of parchment coffee, 80 kg of natural dry or 46 kg of green coffee.

Table 1. Characteristics of coffee per producing state in Mexico.

State	Land with coffee (ha)	Land harvested (ha)	Production (tons of cherry coffee)	Yield (tons of cherry coffee per ha)	Farmers	Average land per farmer (ha)
Chiapas	253,462	251,951	529,250	2.10	174,571	1.45
Oaxaca	185,187	160,888	165,829	1.03	102,513	1.81
Veracruz	153,435	152,450	332,598	2.18	86,961	1.76
Puebla	88,577	70,066	259,246	3.70	47,124	1.88
Guerrero	54,328	53,917	51,152	0.95	21,326	2.55
Hidalgo	26,335	26,335	40,197	1.53	34,616	0.76
San Luis Potosi	22,539	22,539	18,688	0.83	17,552	1.28
Nayarit	19,473	19,473	29,394	1.51	5,401	3.61
Jalisco	4,497	3,984	4,357	1.09	1,106	4.07
Colima	3,018	2,633	2,566	0.97	820	3.68
Tabasco	1,040	1,040	619	0.60	1,211	0.86
Querétaro	300	300	150	0.50	296	1.01
Total	812,191	765,576	1,434,046	1.87	493,497	1.65

Source: Own elaboration with information from SIAP (2008) and (2010).
Note: Land with coffee, farmers, and average land per farmer referred to 2008, while the other data are from the 2008-09 harvesting season.

and sold by intermediaries. Then, we calculated the difference between buying and selling prices and we ended up with (standardized) gross marketing margins.

In the model specification section, we try to translate the theory into an empirical model that we can estimate. The fact that traders decide to have contracts with their buyers is important in understanding the setting in which they interact with each other. Knowing which factors can affect this decision can help us comprehend how the Mexican coffee chain is shaped and how the institutional environment (the rules of the game) relates to trader's decisions to have contracts – the play of the game.

DATA AND METHODS

In Mexico, coffee is produced in twelve states (see Table 1). These states comprise 52 regions, which represent 541 municipalities (Escamilla*et al.*, 2005; SIAP, 2008). For the purpose of this work, two states (Oaxaca and Veracruz) were selected to investigate the relationships among coffee traders and the environment in which they interact. Oaxaca is located on the Pacific Ocean slope, whereas Veracruz is located on the Gulf of Mexico slope. The Oaxaca producing regions are characterized by long periods of dry and hot weather, which helps the picking and processing of coffee beans. As a consequence, most of the coffee that is marketed in this state is either parchment or green coffee. The Veracruz producing areas have humid weather and several periods of rain throughout the year. This means that farmers cannot dry their coffee as easily as in Oaxaca, which results in them selling cherry coffee to the intermediaries (FIRA, 2003).Cherry coffee is highly perishable and preferably must be processed within 24 hours after picking. These differences have an impact on the way in which coffee

traders interact with each other, on their performance, and on the type of mechanisms they choose to enforce their relationships.

After choosing the states of Oaxaca and Veracruz, the next step was to pick the regions (composed of some neighboring municipalities) in which the investigation was to be carried out. In this study, four regions were selected, two in each state. For this purpose we established five criteria: the average altitude of the coffee plantations, having electricity in the households (a proxy for remoteness or isolation), having paved road (a proxy for accessibility to urban areas), coffee cooperative participation, and the number of intermediaries registered in the municipality.

Regarding the intermediaries in the municipality, their number was calculated using the list of coffee traders who were registered with the Mexican Association of the Coffee Production Chain (AMECAFE, for its initials in Spanish) (AMECAFE, 2009). Municipalities with less than four registered intermediaries were considered as suffering from market restrictions. Therefore, municipalities with less than or equal to three intermediaries registered were considered as being faced with high market restrictions. Those municipalities where there were more than three intermediaries were considered as having few market restrictions to commercialize coffee.

With the above criteria, we then classified municipalities in Oaxaca and Veracruz into two categories, based on the number of restrictions they face. Those with four or more high market restrictions, including a low level of intermediary concentration, were grouped in the first category. Those with less than four restrictions but with a high level of intermediaries registered were included in the second category. A randomized selection was then

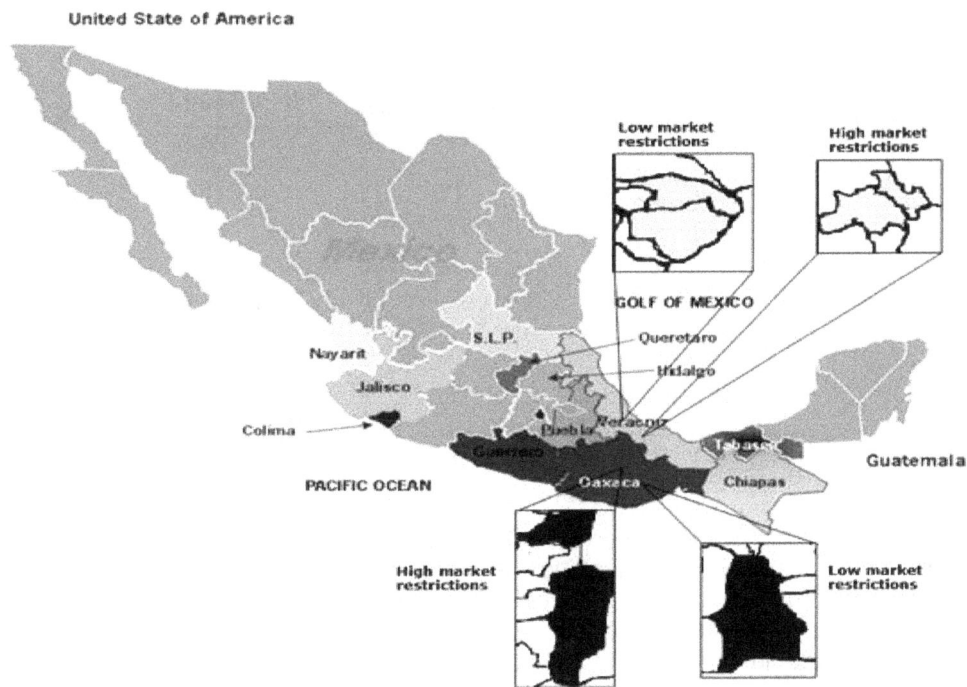

Figure 3. Location of coffee producing states and selected regions in Oaxaca and Veracruz, Mexico.

performed to determine one region per category and state. The four selected regions, two in each state, are shown in Figure 3.

The selected region in Oaxaca with high market restrictions comprises the municipalities of San Felipe Usila and San Felipe Jalapa de Diaz. For the region with low market restrictions in this state, the municipalities included were Pluma Hidalgo, San Pedro Pochutla and Candelaria Loxicha. In the state of Veracruz, the municipalities included in the high market restrictions region were Chocaman and Tomatlan. For the low market restrictions region, the municipalities were Teocelo and Cosautlan de Carvajal.

The second step involved was to ask municipal and local authorities about coffee intermediaries working there. This procedure was undertaken in the field as we visited each of the selected municipalities. We then compared the names given with the ones in the registered list and the ones we found to be non-registered. After having gathered the whole list of intermediaries per municipality, we proceeded to select a random sample from them. Both the registered and the non-registered intermediaries selected added up to a total of 34.

The third step was to include relevant intermediaries that were referred to by the people we interviewed in the field. These relevant traders were described as being one step further in the coffee chain compared to those who referred them. Also, some of the relevant traders operate at a state or even a national level. In total, we surveyed 53 intermediaries. Apart from the surveys with intermediaries, we also held meetings with local

authorities and people involved in the National Coffee System Committees (NCSC).[2] For this purpose, we used semi-structured questionnaires.

It is important to clarify that the traders in our sample ranged from small collectors to exporters, which means that the stage in which each agent operated may vary. Some of them buy directly from coffee farmers (or are in fact coffee growers themselves) and perform their activities at a local level. Others buy coffee from these local intermediaries and sell it to traders who operate at a state or national level (see Annexes 1 and 2). We also interviewed roasters who sell their coffee directly to consumers. These characteristics will be taken into account when we analyze the contracting decisions these agents made. Of course, a contract is an agreement between two parties, so the characteristics of one side are just half of the story to explain contractual choices. However, having these different agents in our sample also allows us to depict how the other side of the contract might behave, since we cover different stages in the supply chain. Unfortunately, because of time and money constraints, we did not interview coffee producers who were the providers of the first stage of the coffee supply chain.

---

[2] The NCSC are non-governmental organisations constituted to serve as mechanisms for permanent planning, communication and consultation between different actors in the coffee chain.

Table 3. Classification of interviewed intermediaries.

Category	Number of respondents	Percentage
Local collectors	18	34.0
Regional collectors	7	13.2
Local cooperatives	4	7.5
Regional cooperatives	3	5.7
Traders/processors	8	15.1
Roasters	3	5.7
Exporters	8	15.1
Others	2	3.8
Total	53	100.0

Source: Own elaboration with data from the survey.

Table 2 shows the general characteristics of interviewed intermediaries. Traders are on average middle-aged, with considerable experience in the business of trading coffee. Most of them started their businesses after the market liberalization reforms took place in the early nineteen nineties.

Another aspect of traders' characteristics was that 43% (23) of the respondents were working in other activities besides coffee marketing. Among these, eleven respondents worked in local grocery stores. The rest of them diversified their income by producing and/or selling other agricultural products and by trading livestock. Even though these intermediaries do not rely solely on coffee as an economic activity, sixteen of them agreed that marketing coffee is the activity that promises them the highest economic benefits. It is important to keep in mind that coffee is a seasonal crop; in the case of Mexico, the harvest season runs from October to April. Respondents devote on average seven months to buy/sell coffee. Those who spend more months in the business and who engage in the processing or roasting of coffee are large exporting companies.

Using data from the survey, we classified the intermediaries by the size of the area where they operate, and also by determining whether they add value to the coffee they buy and sell by processing. As can be seen in Table 3, almost half of the interviewed people operated as collectors on a local or regional basis. This means that this group of intermediaries did not add any value to the coffee they marketed, apart from transport and selection. Only 13% of the respondents were affiliated with a local or regional cooperative. There were only three cases in the sample in which cooperatives directly exported their coffee. The rest of the cooperatives only collected coffee from its members and then sold it to other intermediaries. In terms of value added, we found that 20% of the respondents could be classified as processors/roasters. Also, if we look at regional differences, given the spatial set-up of our field work, we observed that processing activities often took place in areas where competition was high (see Annexes 1 and 2).

Regarding intermediaries who are involved in processing activities, we found that 20% of all respondents were involved in wet processing; all wet processing plants were located in the state of Veracruz, as expected. The average wet processing capacity was 120 quintals per day.[3] Survey data also show that intermediaries in Oaxaca only traded parchment coffee.[4] This proves that our assumptions about differences in the type of coffee that is traded in each state are reflected in the characteristics of our sample. We observed that 30% of the respondents were involved in dry processing activities, with an average capacity of 140 quintals per day.[5] More than half of the intermediaries involved in dry processing were exporting their coffee, and they accounted for all respondents who were classified as exporters. This is understandable, as most of the coffee is exported as green – and therefore processed – coffee.

Three of the respondents were roasted. They operated mainly at a local level. They supplied to local cafeterias or stores, and some even had their own coffee bar. All roasters were located in regions with few market restrictions, which again showed that all value-adding activities were concentrated in areas where it was easier to buy and sell this product.

[3] Coffee processed by the wet method is called wet processed or washed coffee. The wet method requires the use of specific equipment and substantial quantities of water. Following the wet process, the fruit covering the seeds/beans is removed before they are dried.

[4] Parchment coffee is obtained after cherry coffee beans go through wet processing. To obtain green coffee, parchment coffee goes through dry processing.

[5] The dry process is the oldest method of processing coffee. The entire cherry after harvest is first cleaned and then placed in the sun to dry on tables or on patios. The dry process is also known as unwashed or as the natural one.

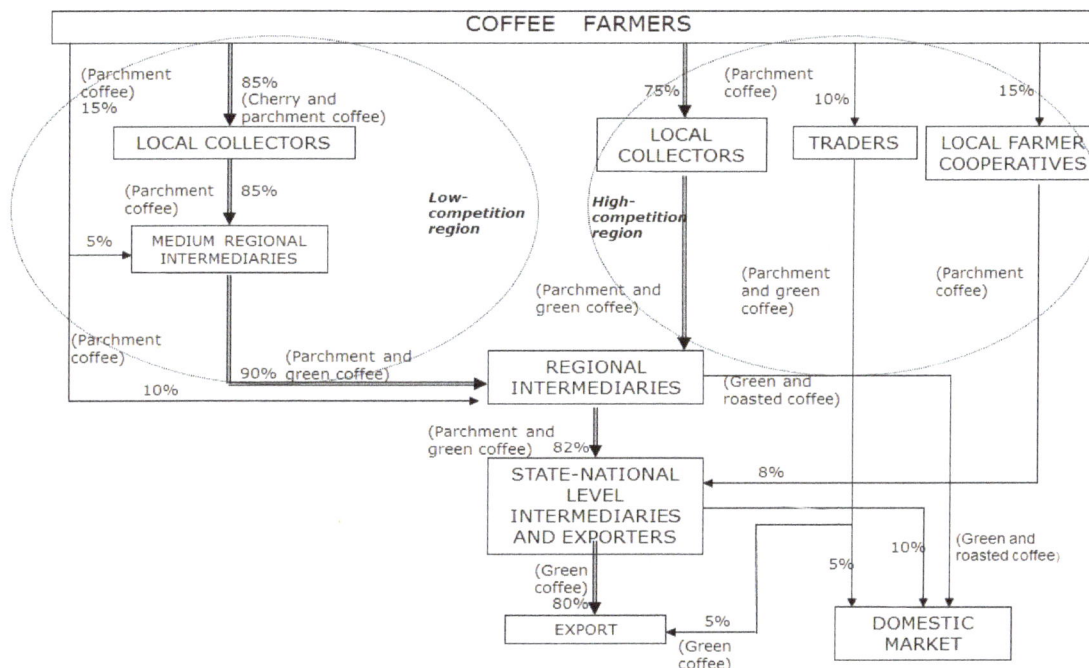

Annex 1. Coffee commercialization scheme in Oaxaca.

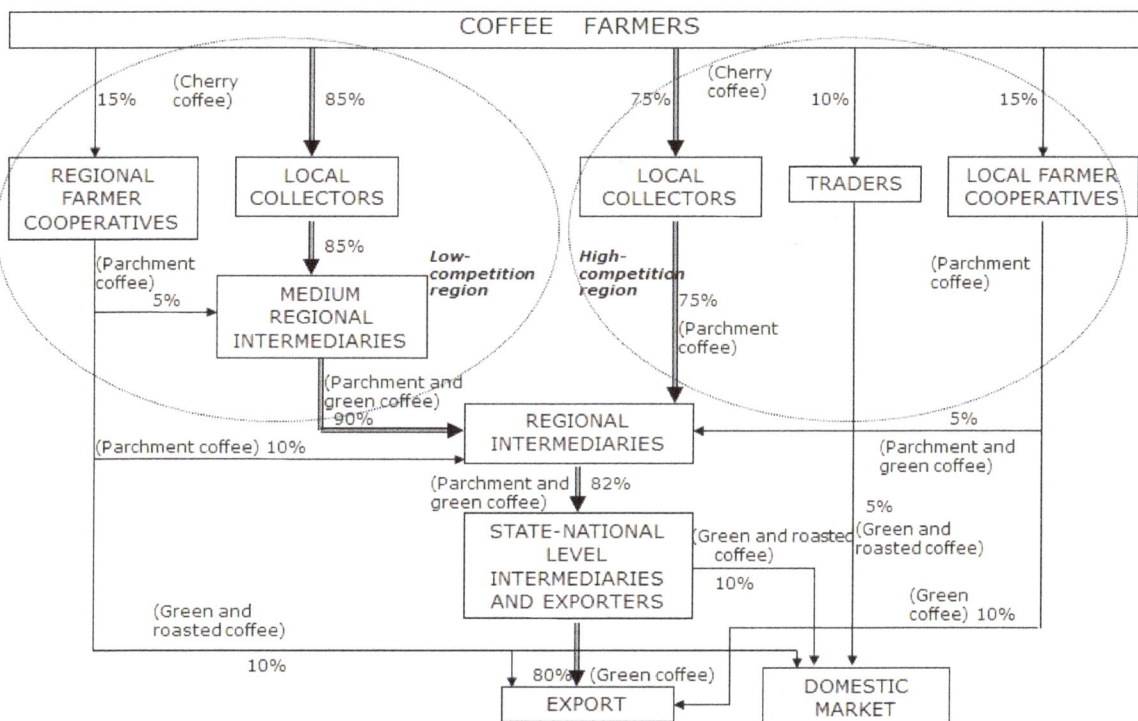

Annex 2. Coffee commercialization scheme in Veracruz.

We also asked traders if they had five or more competitors in the buying area where they operated. A total of 50% of local collectors and 86% of regional collectors mentioned that they had five or more competitors. Also 50% of both local and regional cooperatives stated that they had five or more competitors in the area where they operated. For processors, roasters and exporters, the percentage of traders who answered that they had five or more competitors was 75%, 67%, and 100%, respectively. We

Table 4. Assets owned by interviewed intermediaries.

Asset	Number of respondents	Percentage
Vehicle	30	56.6
Store	43	81.1
Coffee-cupping laboratory	6	11.3
Processing plant	18	34.0

Source: Own elaboration with data from the survey.

can see that in all stages of the supply chain, traders perceived their operational context as a quite competitive environment in terms of the number of traders they observed. However, it is clear that most of our sample consisted of traders present in the primary collection level of the coffee sector. As one of the interviewed coffee producers' leaders affirmed, high competition is established in areas where coffee farmers sell cherry coffee.

In the survey, 94% of the respondents had coffee growers as their suppliers. In addition, some of the interviewees made direct purchases from coffee producers, even when they worked at the regional or national level. Only 28% of the interviewed intermediaries had a contract with their suppliers. Half of those who had agreements stated that it was a verbal one.

An agreement between intermediaries and coffee growers can be beneficial for both participants. Intermediaries provide financial support to coffee producers to implement the picking, mainly to cover the transport of the coffee pickers and payments during the first days or week of the harvesting season. As was commented by some interviewers, they give money in advance to some coffee farmers in order to engage them for the coming harvesting season. Some respondents said that using this kind of financial agreement allowed them to pay a lower price and still receive a large quantity of coffee. The type of agreement set up with the buyers depends on the type of coffee that growers normally deliver.

To have as much coffee as possible, large firms instruct their commissioners to set up some kind of agreement with coffee producers. Normally these commissioners have good knowledge about the type of coffee that can be bought in the region. They are financed with money from the firms in order to fund some coffee producers who will later become their providers. It was found that some companies also had skilled commissioners to provide technical assistance to coffee growers in producing and processing coffee. They did so to improve the coffee quality.

For many small-scale producers, receiving economic support from the buyers establishes the relationship between the growers and the intermediaries. This kind of financing agreements became more common practice under the actual condition than they had been in the ICAs' era; nowadays, no formal bank credit is available for small- and medium-scale farmers in the Mexican coffee sector. For some coffee farmers, an option to tackle this difficulty is to become a member of a cooperative that sometimes has access to economic support from governmental or non-governmental organizations.

Another aspect we investigated in the survey was the ownership of assets by intermediaries, which is shown in Table 4. Looking at regional differences, we observed that there are more intermediaries with vehicles in the zone with high market restrictions in Oaxaca. We argue that this happens because this area is isolated, since it is located in a mountainous area and has a precarious road infrastructure. Thus, intermediaries should own vehicles to overcome these difficulties. This can also imply that traders located in a region with high market restrictions face higher transaction costs due to the mentioned isolation that may reflect in their performance and contract choices. Only few intermediaries own a coffee-cupping laboratory, and they are exporters. Cupping is important since it is a way of inspecting and ensuring the intrinsic quality of the coffee before it is exported.

A third of the interviewed intermediaries own a processing plant, and most of them are located in the state of Veracruz. The fact that coffee growers mainly sell cherry coffee in the Veracruz regions creates a need for processing infrastructure, given that the coffee quality decreases if it is not processed within 24 hours after picking. Related to this, 80% of the respondents own a specific depot or place where they buy and sell coffee. The rest buys and stores coffee either in their house or in the processing plants. Despite this, traders stated that they do not store coffee seeking for better prices. They only use these depots to collect the required amount of coffee for transport later on. Local and regional intermediaries in Oaxaca usually take one week to gather the amount of parchment coffee necessary for a load. In Veracruz, those who buy and sell cherry coffee take one day to collect the coffee and they all sell it within the next 24 hours. Exporters are the ones who take more than a week to gather the amount of coffee needed, given the volume they trade.

When asked about the problems they face as coffee traders, 60% of the respondents mentioned that price volatility was the main obstacle to commercialize coffee. Others mentioned a lack of security when transporting both coffee and money, and quality deficiency as problems. In terms of risk coping strategies adopted by intermediaries, we found that 30% of the sample tried to obtain good information about market conditions and prices. The most common way to gather information about prices was by phone, accounting for 95% of the respondents; internet was used by 37% of the sample. Most local and regional intermediaries made daily phone calls to their buyers to know the price for the coffee they were going to sell.

The most common type of the agreement between intermediaries was an oral contract. Only 15 respondents mentioned they used written contracts with their buyers to reduce any risk and uncertainty, and they were mainly exporters who needed a contract to export their coffee. In these cases, the contract was compulsory to deal with customs authorities. This group of respondents was also the only ones to state that they used the futures market as a price risk management activity. Even though the government is trying to attract small intermediaries and growers into price hedging programs, requirements are not easy to comply with.

Most local and regional intermediaries said they interact with only one downstream trader each year. This shows that repeated transactions take place between traders along the coffee chain. These repeated interactions may also be important when analyzing the (non-)existence of contracts. The fact that traders have known each other for years and that they trust each other may explain the absence of written contracts. Verbal agreements arise in these relationships, as we observed in the survey. However, if there are no written contracts, how can buyers be sure that sellers are going to give them their coffee year after year (or even day after day)? Forty per cent of the sample mentioned that they receive credit from their buyers at zero interest, and all of them stated they use this money to buy coffee.

These repeated transactions between traders are a sign of an environment in which intermediaries are not free to choose who to sell to, especially in the areas faced with high market restrictions. To assess this, we asked intermediaries whether they were now selling to a different person than they were five years ago. We observed that in the Oaxaca region with market restrictions, 60% of the intermediaries had changed buyer, because they were either looking for better prices or the last buyer did not respect their previous agreements. This result can be seen as an indication of intermediaries being free to choose the buyer that best fits their needs.

Not all interviewed traders followed regulations established in the coffee sector. Twenty five per cent of the sample was registered in the National Coffee System (NCS). After registration, agents have to fulfill some requirements and pay an annual fee to be part of this system. The requirements are that one has to be legally established as a firm or, in the case of a natural person,

one needs to have proper identification, a fiscal address, one has to be up to date with tax payments, and one needs to have a written recommendation from an active member of the coffee sector, among other things. Sixty per cent of the respondents were not registered in the NCS as intermediaries.

Most of the intermediaries who were not registered in the system belonged to the categories of local and regional collectors. When asked about the reasons for not being registered they mentioned that there were many requirements and a lot of paper work in order to be in the system. Since registered traders issue a bill to coffee growers in which the quantity and price of the transaction is stated, and this bill is then used by growers to get government subsidies, most unregistered intermediaries get these bills from buyers working downstream in the coffee chain to satisfy the producers' demand for the bills.

The coffee chain starts with cherry coffee sold by growers; it is then transformed into parchment coffee, which is processed into green coffee, the type that is commonly exported. After that the green coffee can be transformed into roasted and ground coffee. In terms of the conversion from cherry to parchment coffee, we know that 245 kilograms of the former make one quintal of the latter. Also, the average cost for doing wet processing is 146 Mexican pesos per quintal. In this case, with the weight conversion, we multiply the price of one kilogram of cherry coffee to get a price per quintal of parchment coffee and add the processing cost to make it comparable.

When going from green to parchment coffee, the weight conversion applied is that a quintal of green coffee is equivalent to 0.80 multiplied by the weight of a quintal of parchment, and the costs to be subtracted are 110 Mexican pesos per quintal, indicating dry processing. We subtract the costs, since green coffee is one step further in the processing stage and to get prices in terms of parchment coffee can be seen as going backwards in this stage. To go from the prices of roasted coffee to those of parchment coffee, and knowing that the former is sold per kilogram, we multiply by 37.5 to get the weight equivalent of a quintal (in kilograms). We also have to subtract the associated processing costs, which in this case are 267 Mexican pesos; this is because we subtract both the costs of going from parchment to green coffee (110 Mexican pesos per quintal) plus the costs of going from green to roasted coffee (157 Mexican pesos per quintal).[6] After this process, we obtained all prices in terms of quintals of parchment coffee purchased and sold by intermediaries. Then, we calculated the difference between buying and selling prices and we ended up with standardized gross marketing margins. From the 53 traders in our sample, 44 observations were affected by this conversion. We could not perform the conversion on

---

[6] The processing costs per each of the steps in the supply chain were obtained from the interviewed processors.

Table 5. Descriptive statistics of variables included in the analysis (N = 53).

Variable	Units	Mean	SD
Having contract	1 if the intermediary has a contract	0.60	0.49
Margins[1]	Mexican pesos per quintal	410	687
Experience[2]	Years of experience in the coffee business	14.09	9.21
Other business	1 if the intermediary has another business	0.43	0.50
Owns vehicle	1 if the intermediary owns a vehicle	0.57	0.50
Restricted region	1 if the intermediary works in a restricted region for marketing coffee	0.40	0.49
Roaster	1 if the intermediary is a roaster	0.23	0.42
Wet processing plant	1 if the intermediary owns a processing plant	0.25	0.43
Dry mill	1 if the intermediary owns a dry mill	0.30	0.46
Long-term relationship	1 if the intermediary established a long-term relationship	0.34	0.48
Sells cherry coffee	1 if the intermediary sells cherry coffee	0.19	0.39
Competition	1 if there is competition in the region where the intermediary works	0.68	0.47
Registered in the NCS	1 if the intermediary is registered in the NCS	0.42	0.50
Buyer is registered	1 if the buyer is registered in the NCS	0.60	0.49
Volume of sales[4]	Quintals per season	4,250	7,708
Veracruz	1 if the intermediary resides in the state of Veracruz	0.49	0.54

Source: Own elaboration with data from the survey.
Notes: [1]The minimum is 25 and the maximum is 3,000.
[2] The minimum is 2 and the maximum is 46.
[3] The other business x owns a vehicle.
[4] The minimum is 50 and the maximum is 36,000.

the whole sample, since some traders refused to tell us their selling price. Furthermore, a limitation to this conversion is that the average costs we took into account in this process could change for each trader given their infrastructure and technology. However, we also could not get more precise costs because most interviewees did not share this information with us.

With the above characterization of coffee traders in the states of Oaxaca and Veracruz, Mexico, we constructed a set of variables to answer the research questions. A logit regression was performed to find variables affecting the intermediaries' decision to have a contract (or not) with their buyer. In this procedure, we used several explanatory variables related to individual, regional and market characteristics to predict the probability of a trader having a contract. To assess the performance of traders in the Mexican coffee sector, we calculated their gross margins. Then we investigated which variables affected the intermediaries' performance thus measured. To do so, we performed an ordinary least squares regression.

CONTRACT CHOICES
Model Specification

The decision to contract with another agent can be expressed a discrete choice model. In this case, an intermediary will choose to enter into a contract with his buyer if the expected benefits of having one are greater than those of arranging the transaction in an alternative way (Masten and Saussier, 2000). In other words, a trader will choose to enter into a contract if the expected utility of having one is greater than the expected utility of not having a contract, otherwise the choice will be not to have a contract.

We constructed a set of variables that represent the socio-economic characteristics of intermediaries, as well as some indicators of their business practices and their relationship both with other traders and the environment in which they perform their activities (Milagrosa, 2007). Table 5 lists the variables that were included in the regressions.

The dependent variable was a dummy that took the value of one if a trader had a contract and zero if not. It also stood for those agents who were part of a vertically integrated organization.

As part of contracting decisions of traders in the Mexican coffee sector, factors like, among other things, asset specificity, bounded rationality, power relationships and opportunistic behavior have an effect on the traders' contracting decision. At the same time, these factors can also have an effect on the performance of the firm.

Table 6. Determinants of contract engagement (logit).

Variables	Coefficient	Marginal effect
Experience	-0.027 (0.582)	
Restricted region	).591 (0.450)	
Roaster	-3.285 (0.016)**	-0.673
Wet processing plant	-2.099 (0.050)**	-0.478
Dry mill	2.657 (0.037)**	0.449
Long-term relationship	1.054 (0.246)	
Sells cherry coffee	-2.701 (0.026)**	-0.587
Veracruz	-0.042 (0.967)	
Intercept	1.462 (0.161)	
Observation numbers	53	
Likelihood Ratio (LR) index	-25.32	
McFadden's R²	0.288	

Note: P-values in parenthesis. * and ** refer to significance at 10% and 5%, respectively.

Results of Contract Choices

Table 6 presents the results of the logit estimation of the model for the decision made by traders to have a contract with their buyer.

The first variable found to be significant (at a 5% level) in our model was the one that indicates whether a trader is a roaster. The magnitude of the marginal effect is also significant, since being a roaster decreases the probability of having a contract by 67%. We expected a positive sign (meaning an increase in the probability of having a contract) given the fact that roasters are traders who are vertically integrated and have made specific investments. Most roasters do not trade large amounts of coffee and usually sell their coffee to incidental customers. This means that they may choose other types of arrangements with these clients, especially spot market transactions, rather than formal contracts.

The wet processing plant variable negatively affects the decision to have a contract. Apparently, wet processing does not make the owner dependent on single buyers, and in this sense, the asset is not 'specific'. The opposite result was found for having a dry mill which is more used further downstream, notably by exporters. This magnitude of the associated marginal effect of the variable of owning a dry mill increases the likelihood of having a contract by 45%.

The last variable that was found to affect the contracting decision is the one depicting a trader who sells cherry coffee. The associated marginal effect shows that selling cherry coffee decreases the probability of having a contract by 59%. This finding is in line with our expectations. Traders who sell cherry coffee face repeated transactions with their buyer, and then trust-like relationships can be chosen over formal contractual agreements (Fafchamps and Minten, 1999).

In our approach of selecting the four regions, we included two regions as restricted in terms of having least quantity

of traders operating, low level of producers' organization and being far from the coffee buying centers. We included this variable in the analysis but found that it was not significant in explaining the traders' contracting. Hence the overall marketing environment is no important factor for contracting.

TRADERS' PERFORMANCE
Model Specification

To find out which variables explain the variation of the margin, the (standardized) gross margins were included in a linear regression model as dependent variable. In this sense, we regressed the margins on a set of variables depicting trader's socio-economic, marketing and institutional characteristics. This linearization allows us to use the concepts drawn from the theory and apply them in an empirical way. In this study, by performance we mean the gross margins traders can get for the coffee they market, that is, the difference between purchase and selling prices.

Experience in marketing coffee was the first independent factor included in the model. As we have mentioned, specific experience in marketing coffee is not the same as experience in the coffee sector, which can be given by a trader's age. Most of the traders in our survey have had experience in the coffee sector even though they were not engaged in trading coffee. We expected a positive relation of this factor with the traders' performance. In terms of the characteristics of the firms, we utilized the variable of whether a trader had another activity to obtain income from. We expected that if traders diversified their livelihood they would have lower margins because they did not rely solely on the profits from trading coffee (Jabbar et al., 2008).

The theory tells us that assets owned by a trader play a role in their performance. This is why we included vehicle ownership as an independent variable. The expectations

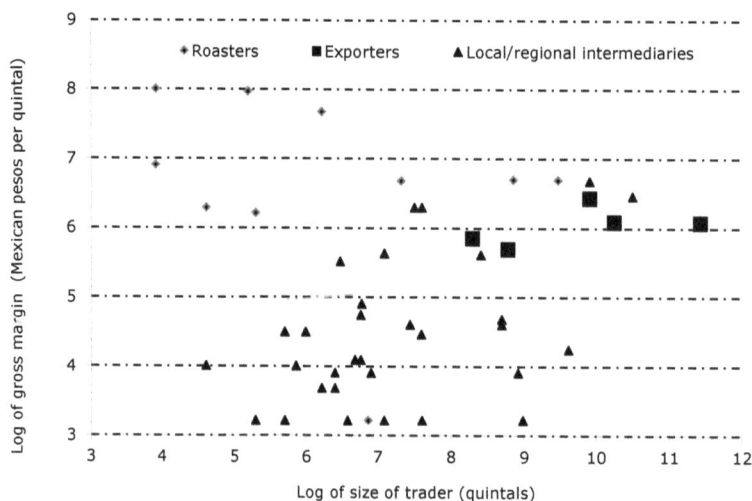

Figure 4. Size of intermediaries vs. marketing margin (own elaboration with data from the survey).

for this variable were twofold. On the one hand, traders who own a vehicle may have more working capital than those who do not. In this sense, margins for these traders may be low, because they have the ability of having a large volume of business (see Figure 4). This means that even though they have a small margin per quintal sold, they have the capital to buy more coffee and compensate (Jabbar et al., 2008). On the other hand, since a vehicle forms a costly asset, this would require a higher return from their business, thus margins might be high.

The next set of variables show first the degree of integration that a given firm can have and secondly, the stage in the coffee chain in which a trader can be found. In our sample, some coffee agents had large margins but small businesses in terms of the quantity of the coffee they market.

We included a variable that indicates whether intermediaries traded cherry coffee. Traders who buy and sell cherry coffee should do the buying and selling transactions within 24 hours. This can be detrimental to their performance, since the urge to sell their coffee can deprive them of the opportunity to find better options in the market. This may also show that, given the environment in which they participate, they decide to sell cherry coffee because this means incurring lower costs. Hence, we expected a negative sign for this variable in the OLS regression.

Operating in a (non-)competitive environment affects the way traders perform. This is why we included a variable that indicates whether a trader operated in a region with more than five traders. We expected that the more competitive the environment was, the smaller the margins would get (Mose, 2007). This means that firms will look to compensate low margins with larger volumes of coffee being marketed.

Being registered in the NCS is another explanatory variable in the OLS model. Traders who had more

knowledge about how to develop their business performed better or had lower margins than those who had less knowledge (who were not registered in the NCS). Intermediaries might give lower prices to their providers in exchange for the bills; thus, traders would seek higher margins.

The next variable included in the model was the decision made by traders to have a contract with their buyer. This variable tries to link the concepts of contract with the traders' performance. Changes in the institutional framework should lead to contractual arrangements between parties in order to minimize costs (North, 1990). This means that having a contract should lead to lower margins as a consequence of such cost minimization. We should keep in mind that the margins we consider for our model consist of the difference between buying and selling, and that the contract we analyze is between traders and their buyers. In this sense, if there is in fact a contractual agreement, it could only secure the selling price. We used the estimated probability of having a contract obtained from the logit model to avoid a potential endogeneity problem.

Results of Traders' Performance

Taking into account the possibility that having a higher gross margin would make it more likely to there would be more money to start another business or to buy a vehicle; and also that having a higher gross margin, traders would be more likely to be registered in the NCS, we suspected that there was an endogeneity problem. We did a Hausman test for those variables to check this and found that the problem was not present.

The results for the second estimation procedure are included in Table 7. The first variable listed as significant for our analysis of margins is the 'sell cherry coffee' dummy. The model shows that a trader who sells cherry

Table 7. Factors explaining gross log-transformed margins (OLS).

Variable	Coefficient
Experience	-0.023 (0.124)
Other business	-0.145 (0.595)
Owns a vehicle	0.086 (0.767)
Roaster	0.853 (0.122)
Dry mill	0.271 (0.568)
Sells cherry coffee	-1.936 (0.000)***
Competition	-0.689 (0.021)**
Registered in the NCS	1.070 (0.007)***
Buyer is registered in the NCS	-0.254 (0.434)
Having a contract (predicted)	-1.599 (0.062)*
Intercept	6.831 (0.000)***
Observation numbers	43
R squared	79.99

Note: P-values in parenthesis. *, ** and *** refer to significance at 10%, 5%, and 1%, respectively.

coffee will have lower margins than one who sells other types of coffee. As we mentioned before, traders dealing with cherry coffee have to buy and sell their product within the next 24 hours to avoid loss of quality. Traders who sell cherry coffee also face lower transaction costs and this is reflected in lower margins. The latter may indicate that there are not enough incentives and market institutions encouraging them to invest in adding processing value to their product.

The results show that an intermediary who operates in a more competitive environment experiences a 69% decrease in the margins. This finding is consistent with the theory in the sense that higher competition lead to a lower marketing margin.

The output of our regression indicates that traders who are registered and arguably operate under the rules set by the current structure of the coffee sector get larger margins than those who are not. This finding may be related to the fact that registered buyers are obliged to declare taxes and pay the government a certain amount for the value of the coffee they buy. The latter is required by a government program named The Stabilization of Coffee Pricing Fund.

The fact that traders decide to engage in contractual agreements with their buyers reduces the gross margins. This shows that, as theory suggested, agents who make the decision to have contracts may face lower costs, since the reason to enter into such contracts was to tackle risks and uncertainty present in the market. In this sense, when dealing with lower costs, traders can settle for smaller per quintal margins. The rest of the variables included in the model made no significant contribution toward explaining coffee traders' performance.

CONCLUSION

Liberalization of the market should bring about a change in the institutional environment of the market. This has to

result in a more competitive framework, in which agents have to find new ways to relate to each other to increase their earnings. In this paper, we investigated whether that proposition can be proven for the Mexican coffee supply chain. Twenty years have passed since the Mexicans experienced a transition from a state-controlled system to a free market. We provide some evidence on the successes and failures of the new institutional framework for this sector.

One of our results indicates that coffee agents have involved themselves in different activities as a response to the transition from a state-led commodity chain to a liberalized environment. We described how some of them decided to vertically integrate in order to reduce transaction costs and achieve higher efficiency in the market, relying on a larger volume of business to have enhanced profits. Others, on the other hand, invested in assets that allowed them to add value to their product and to attain a better performance in terms of margins per unit of product sold.

Most of the traders are unknown to the government, and thus are unable to participate in any governmental initiative. Data indicate that 60% of the traders are not registered in the National Coffee System. That situation creates unsettling circumstances for small-scale traders.

Traders having processing facilities do not have more contracts. If agents are vertically integrated, such as roasters, they have less contracts, but if the traders have processing facilities in an earlier stage of the supply chain such as dry mills, they have more often contracts. In the first stage of the supply chain, having wet mills reduces the probability of having contract. These results are in line with the transaction costs economic theories in the sense that most contracting occurs where the market is thinnest: in the middle stage of the chain.

An interesting result is that high degree of competition will increase the chain's efficiency and lower the margins. Having a contract helps in reducing the uncertainty and any opportunistic behavior, hence an increase in the performance (lower margin) is observed.

An important finding is that the stage of the Mexican coffee sector we analyzed appears to be reasonably competitive. A challenge for further research related to the issues we tackle in this research is to gather precise data to perform a similar analysis at all stages of the coffee supply chain.

REFERENCES

Akiyama T (2001). Coffee market liberalization since 1990,in Baffes J, Larson DF, Varangis P (Eds.)*Commodity market reforms: lessons of two decades*. World Bank Publications, pp. 83-120.

Asociación Mexicana de la Cadena Productiva Café (AMECAFE) (2009). Lista de intermediarios registrados en el sistema informático de la cafeticultura. México, D.F.

Escamilla PE, Ruiz O, Díaz G, Landeros C, Platas R, Zamarripa C (2005). El agroecosistema café orgánico en México. *Manejo Integrado de Plagas y Agroecología*, Costa Rica, 76, 5-16.

Fafchamps M (2002). Spontaneous market emergence. *Topics in Theoretical Economics*, 2(1): 50.

Fafchamps M (2004). Market institutions in sub-Saharan Africa: Theory and evidence.*MIT Press*.

Fafchamps M, Minten B (1999). Relationships and traders in Madagascar. *Journal of Development Studies*, 35(6), 1-35.

Fafchamps M, Gabre-Madhin E, Minten B (2005). Increasing returns and market efficiency in agricultural trade.*Journal of DevelopmentEconomics*,78(2): 406-442.

Fideicomisos Instituidos en Relación a la Agricultura (FIRA) (2003).*Situación de la Red Café: Oportunidades de Desarrollo en México*. Morelia, Michoacán, México.

Fischer C, Hartmann M, Reynolds N, Leat P, Revoredo-Giha C, Henchion M (2009). Factors influencing contractual choice and sustainable relationships in European agri-food supply chains. *Europe Review of Agricultural Economic*, 36(4), 541-569.

Gabre-Madhin E (2001). *Market Institutions, Transaction Costs, and Social Capital in the Ethiopian Grain Market*. International Food Policy Research Institute.

Gabre-Madhin E (1999). *Of Markets and Middlemen: The Role of Brokers in Ethiopia*. International Food Policy Institute.

Gilbert C 1987). International commodity agreements: design and performance. *World Development*, 15(5), 591-616.

Jabbar M, Benin S, Gabre-Madhin E, Paulos Z (2008). Market institutions and transaction costs influencing trader performance in live animal marketing in rural Ethiopian markets. *Journal of African Economic*, 18.

Janvry AD, Fafchamps M, Sadoulet E (1991). Peasant Household Behaviour with Missing Markets: Some

Paradoxes Explained. *The Economic Journal*, 101(409), 1400-1417.

Key N, Sadoulet E, De Janvry A (2000).Transactions costs and agricultural household supply response. *American Journal of Agricultural Economics*, 82(2), 245-259.

Masten S, Saussier S (2000). Econometrics of contracts: an assessment of developments in the empirical literature on contracting. *Revue d'économieindustrielle*, 92(1): 215-236.

Mehta A and Chavas JP (2008). Responding to the coffee crisis: what can we learn from price dynamics? *Journal of Development Economics*, 85(1-2): 282-311.

Milagrosa A (2007). Institutional economic analysis of vegetable production and marketing in Northern Philippines: social capital, institutions and governance. Wageningen University, Wageningen, the Netherlands.

Mose LO (2007). Who gains, who loses? The impact of market liberalisation on rural households in Nortwestern Kenya. Wageningen University, Wageningen, the Netherlands.

North DC (1989). Institutions and economic growth: an historical introduction. *World Development*, 17(9), 1319-1332.

North DC (1990). *Institutions, Institutional Change and Economic Performance*. Cambridge: Cambridge University.

Pérez GV, Cervantes E, Burstein J (2001).Case study of the coffee sector in México. Medford, Mass. Tufts University. Global Development and Environment Institute.

Porter ME (1998). Competitive strategy: techniques for analyzing industries and competitors: with a new introduction. New York. USA: Free press.

Ruben R, Boselie D and Lu HL (2007a). Vegetables procurement by Asian supermarkets: a transaction cost approach. *Supply Chain Management*, 12(1), 60-68.

Ruben R, van Boekel M, van Tilburg A, Trienekens J(2007b). *Tropical Food Chains: Governance Regimes for Quality Management*. Wageningen Academic Publishers. Wageningen, the Netherlands.

Santoyo CVH, Díaz Cárdenas S, Rodríguez Padrón B and Pérez PérezJR(1994). *Sistema Agroindustrial Café en México: Diagnóstico, Problemática y Alternativas*. Universidad Autónoma Chapingo. Chapingo Estado de México.

Schroeter J, Azzam A (1991). Marketing margins, market power, and price uncertainty. *American Journal of Agricultural Economics*, 11.

Sexton RJ, Lavoie N (2001). Food processing and distribution: an industrial organization approach. Agricultural Economics, 1, 863-932.

Shervani TA, Frazier G, Challagalla G (2007). The moderating influence of firm market power on the transaction cost economics model: an empirical test in a forward channel integration context. *Strategic Management Journal*, 28(6), 635-652.

Sistema de Información Agrícola y Pecuaria (SIAP) (2008). Base de datos del censo nacional de productores de café, Actualizado a julio de 2008. México, D.F.

Sistema de Información Agrícola y Pecuaria (SIAP) (2010). Sistema Producto Café. Actualizado a julio de 2010.

Snyder R (1999). After neoliberalism: the politics of deregulation in México. *World Politics*, 51(2), 173-204.

Williamson OE (1979). Transaction cost economics: the governance of contractual relations. *The journal of Law and Economics*, 22, 233-261.

Williamson OE (2000). The new institutional economics: taking stock, looking ahead. *Journal of Economic Literature*, 38(3), 595-613.

Winter-Nelson A, Temu A (2002). Institutional adjustment and transaction costs: product and inputs markets in the Tanzanian coffee system. *World Development*, 30(4), 561-574.

15

Marketing volume transaction analysis of dates in Saudi Arabia

S.H. Alkahtani[1*], A. M. Al-Abdulkader[2], S.M. Ismaiel[1]

[1*] King Saud University, Saudi Arabia.
[2] King Abdulaziz City for Science and Technology, Riyadh 11451, Saudi Arabia.
*Corresponding author email: safark@ksu.edu.sa

Marketing volume transaction in Saudi Arabia, including marketing characteristics, volume, prices, marketing channels, and loss analysis, were considered as study objectives. Results of the research paper showed that the average annual activity in date marketing amounted to about 820.5 tons of date as a total transaction for an average dealer. This amount contains all date varieties in the major date production regions including Sukkari, Khalas, Segae, Barhi, Ajwa, and other date varieties. The average annual activity in date marketing for a wholesaler was amounted to about 611.40 tons of date as a total transaction from the different purchasing sources with an average price of about SR 13471.7/ ton. The average annual activity in date marketing for a retailer was amounted to about 308.70 tons of date as a total transaction from the different procurement sources with an average price of about SR 16095.90/ ton. Waste was one of the important key inductor to improve marketing efficiency in general. Al-Madinah Al-Munawara represented the highest date marketing waste out of date marketing purchasing (5.8 per cent). For all regions, the date marketing waste out of marketing purchasing was 4.6 per cent.

Key words: Marketing volume transaction, marketing channels, date loss.

INTRODUCTION

Dates has a special status in the economic structure of Saudi Arabia. Saudi Arabia produces more than 450ofdates varieties from more than 25.1million date palm trees (MOA, 2014). Known varieties and consumers preference are different from one region to another and from one consumer to another. Sukkariis the most famous date variety in Al-Qassim region, Khalas is the most famous date variety in Al-hasa region, while Ajwa is the most famous date variety in Al-Madinah Al-Munawara region.

The production of dates increased from about 350 thousand tons in 1980 to about 1.03 million tons in 2012, with an increase equivalent to about three folds. This is of course due to the increase in cultivated area of palm trees from about 60, 4 thousand hectares in 1980 to about 160 thousands hectares in 2012 with an increase

by more than two folds during the same period. Accordingly, the productivity increased by about 16 per cent, from 5.67 ton ha^{-1} in 1980 to about 6.56 ton ha^{-1} in 2012 which could be attributed, also, to the application of the advanced technologies and practices in date farming (FAO, 2014).

Date marketing outputs transaction passes through two main marketing channels in Saudi Arabia, traditional marketing channel from producer to consumer without passing through date processing plants, and marketing through date processing plants. Producers under traditional marketing channel sell their produced dates directly at the local markets and at neighboring markets without sorting and grading, fumigation and washing, and regularly, sold as fresh or dried dates. However, date marketing through processing plants is characterized with better quality than dates sold via traditional date marketing channel, where the producers of dates in production areas sell their production to date processing plants located in their areas, according to the quality specifications determined by the plants. The price of dates is determined on the basis of date quality and quantities displayed (Alkahtani el al, 2011; Alshuaibi, 2011).

Date manufacturing and processing are encouraged in Saudi Arabia, to increasing the added value of date production, to provide more job opportunities, to solve a lot of the marketing problems faced by date producers, to achieve a balance between supply and demand, and to find a rewarding and stable prices for producers and favorable prices for consumers.

Date prices are one of the key market signals to producers, where a number of important decisions are taken to invest in the date sector, as well as the consumer decision is made in accordance to the suitability of these prices to their livings. Date prices are usually influenced by the prevailed market powers of demand and supply (John,1991).

Marketing outputs refers to low dates producers share from the price paid by the consumer at the same time marketing margins absorb the largest share (Alkahtani and Elfeel 2006; Alkahtani el al, 2007). Some farmers attempt to perform marketing functions in order to convert a portion of the intermediaries profits to their favor, yet, it did not come to their desired benefits. Thus, This paper aims to explore marketing output transaction in Saudi Arabia, including marketing characteristics, volume, prices, and date loss analysis.

METHODOLOGY

The data required for the research project was collected through a questionnaire prepared to survey date marketing units (DMU's)at the selected date market places in Saudi Arabia, namely, Al-Madinah Al-Munawara, Riyadh, Al-Qassim, and Al-hasa regions (Study Areas). The questionnaire includes general and specific information about marketing characteristics, channels, volume, and loss. The qualitative analysis in the research project includes measuring the mean, the standard deviations, duplicates and percentages of the major variables to identify personal and functional characteristics of the study sample and responses to questionnaire questions. The analysis of variance (One-way ANOVA) was utilized to identify differences between the responses of the respondents to assess the level of significant differences when variables are composed of three categories and more(Dickey and Fuller, 1979). Another qualitative analysis included SCHEFFE Test to validate differences if there are significant differences by using analysis of variance. T-test was utilized to indicate the differences between the responses of the

respondents, and to assess significant differences when the variable is composed of two categories only (Engle el al, 1987; Freedman, 2009).

RESULTS AND DISCUSSION

Marketing Characteristics

The study sample included 298 DMU's. Table (1) presents the basic characteristics of the study sample. Table (1) showed that study areas representation in sample study goes with the importance of date marketing places all over Saudi Arabia, as 33.9 per cent of interviewed DMU's had been done in Al-Madinah Al-Munawara which was considered the most important date marketing region for Saudi dates and combines all types of DMU's (retailers, wholesalers, and exporters). The study sample contains all trade types, where retailing represents about 43 per cent of the total study sample. Also, Al-Madinah Al-Munawara procures dates from all varieties and locations all over Saudi Arabia for supplying Saudi dates for national and international date consumers. Most of the sampled DMU's were specialized only in date marketing (72.5 per cent). While most of non-specialized DMU's (about 87 per cent) practice agricultural marketing activities. A very limited portion of DMU's were highly educated (about 8 per cent) while more than 90 per cent of them are secondary educated or lower.

Dates Importance in DMUs Activities

Table (2) illustrated the importance of dates in dealer's activities. The minimum dates volume to other commercial activities for the sample is 15 ton while the maximum was 100 ton. Years of experience ranged from one year as minimum to 50 years as maximum, with an average of 12 years. This indicated that the studysample gives great confidence to the accuracy of the results, in prospective of the decisions and plans to the development of marketing efficiency. The importance of dates activities as an annual income to DMU's are distributed from 10 per cent to 100 per cent

Marketing outputs

Dates Sources and Prices

Average annual activity in date marketing amounted to about 820.5 tons of date as a total transaction for an average DMU's. This amount contains all date varieties including Sukkari, Khalas, Segae, Barhi, Ajwa, and other date varieties, respectively. The purchasing sources contained owned date farms, direct purchase from other date farms, buying through auction transaction and

Table 1. Basic Characteristics of the Study Sample of Date Marketing Unites (DMU) in Saudi Arabia – 2012.

Variable	Frequency	Per cent
Study areas		
Riyadh	75	25.2
Al-Madinah Al-Munawara	101	33.9
Al-hasa	49	16.4
Al-Qassim	73	24.5
Marketing Channels		
Wholesaling	47	15.8
Retailing	129	43.3
Wholesaling and Retailing	106	35.6
Retailing and Exporting	3	1.0
Wholesaling, Retailing and Exporting	6	2.0
Undefined	7	2.3
Specialization in date marketing activities		
Specialized	216	72.5
Non-Specialized	82	27.5
Other activities (for non-specialized)		
Agricultural Marketing Activity	71	86.59
Non-Agricultural Marketing Activity	11	13.41
Dealer background		
Inherited business	116	38.6
Un-inherited business	182	61.1
Educational level of dealer		
Lower than hi-school	150	50.3
Hi-school	120	40.3
Higher Education	24	8.1
Undefined	4	1.3
Scale of trade (ton)		
Small scale	<10	17.2
Medium scale	10- <50	41.9
Above medium scale	50 – <100	11.0
Large Scale	100 – 500	22.5
X-large Scale	>500	7.4

Source: Study Sample, 2012.

contracting. Generally speaking those sources represent respectively.

Average date purchasing prices amounted to about SR 17142/ton. The price varies according to date variety and also according to purchasing sources, (Table 3).

Wholesaling Sources and Prices

The average annual activity in date marketing for a wholesaler was amounted to about 611.40 tons of date as a total transaction from the different purchasing sources with an average price of about SR 13471.7/ ton (Table 4). Owned farms represented the highest source of dates marketing with around 216 ton. In terms of prices prospective, contracting method represent the lowest prices over all the different dates varieties (3277 SR/ton).

Contracting approach was the best procuring source, so that; it is recommended to increase relying on this source in order to improve marketing efficiency.

Retailing Sources and Prices

The average annual activity in date marketing for a retailer was amounted to about 308.70 tons of date as a total transaction from the different procurement sources with an average price of about SR 16095.90/ ton (Table 5).

Impact of Explanatory Factors on Date Marketing Outputs

Table (6) summarized descriptive analysis for the main

Table 2. Descriptive Analysis of the Importance of Date in Date Marketing Unites (DMU) Activities in Saudi Arabia - 2012.

Variable	Min.	Max.	AVR.	SDV.	C.V
Riyadh					
Relative importance of dates in commercial activities (per cent)	40	100	93.7	15.2	16.2
Years of Experience (year)	1	50	11.9	10.4	87.8
Relative importance of dates in annual income (per cent)	15	100	76.9	23	30
Al-Madinah Al-Munawara					
Relative importance of dates in commercial activities (per cent)	15	100	90.5	23	25.5
Years of Experience (year)	2	45	11.3	7.6	67.2
Relative importance of dates in annual income (per cent)	10	100	82.1	24.2	29.2
Al-hasa					
Relative importance of dates in commercial activities (per cent)	40	100	88.5	17.3	19.5
Years of Experience (year)	5	35	13.9	7.3	52.9
Relative importance of dates in annual income (per cent)	30	100	76	18.1	23.8
Al-Qassim					
Relative importance of dates in commercial activities (per cent)	60	100	94	10.3	10.9
Years of Experience (year)	1	30	12	5.9	49.3
Relative importance of dates in annual income (per cent)	40	100	82.1	15.2	18.5
All Study Areas					
Relative importance of dates in commercial activities (per cent)	15	100	91.9	17.7	19.3
Years of Experience (year)	1	50	12	8	66.7
Relative importance of dates in annual income (per cent)	10	100	80	21.1	26.4

Source: Study Sample, 2012.

Table 3. Average Date Purchasing Sources and Prices for Selected Date Varieties in Saudi Arabia Per Dealer

Date Variety	Owned Farms		Other Farms		Auction		Contracting		All Sources*	
	Ton	SR/ton	Ton	SR/ton	Ton	SR/ton	ton	SR/ton	Ton	SR/ton
Sukkari	45.9	18324.1	34.6	20544.8	23.3	18282.9	26.9	9525.0	130.7	17093.2
Khalas	46.8	11563.9	110.3	10556.8	93.3	13880.8	51.7	8060.0	302.1	11312.3
Segae	45.1	13133.3	24.8	10984.6	15.7	15596.4	15.4	14750.0	101	13234.8
Barhi	14.1	7714.30	17.3	10140.6	6.4	10155.7	4.0	11250.0	41.8	9429.5
Ajwa	69.3	31166.7	9.8	48000.0	28.1	51837.2	50.0	32000.0	157.2	36173.2
Other	26.4	9006.50	16.8	10415.2	39.0	13772.9	5.5	8548.4	87.7	11367.2
Total *	247.6	18094.6	213.6	13888.8	205.8	19557.6	153.5	16886.9	820.5	17141.7

Source: Study Sample, 2012.
*quantity in total tons, prices in weighted average

date marketing outputs, including, quantities, prices, and added values, followed by a spill out discussion on the impact of some key factors affecting the date marketing outputs, including, marketing channels, geographic factor (study area), educational level of DMU, concentration of date marketing, other activity besides date marketing, and length of experience in date marketing.

Type of Marketing Channel

Table (7) showed that average date marketing margin doesn't show significant variation according to various marketing channels. However date annual quantity transaction show wide range of variations among various marketing channel. Date marketing units which practice wholesaling, retailing and exporting activities perform significantly more quantity of transaction. It was clear from the table that retailers (whether or not they practice date wholesaling activity) usually purchase date with significantly higher price level than wholesaler (whether wholesaler export or retail date or not). This result confirm the three main price level of date marketing system, farm gate price, wholesale price and retail level. Selling price showed the same pattern in differences

Table 4. Wholesalers Date Purchasing Sources and Prices for thy Selected Date Varieties in Saudi Arabia

Date Varity	Owned Farm		Other Farms		auction		Contracting		All Sources	
	Ton	SR/ton	Ton	SR/ton	Ton	SR/ton	Ton	SR/ton	Ton	SR/ton
Sukkari	14.0	13300.0	10.6	32800.0	40.3	14400.0	.5	5625.0	65.4	17079.7
Khalas	62.9	8738.1	41.3	10250.0	58.4	10642.9	100.3	3250.0	262.9	7305.3
Segae	30.3	10285.7	19.3	9766.7	11.7	12861.1	.5	9000.0	61.8	10599.2
Barhi	17.0	4375.0	-	-	14.0	5250.0	-	-	31	4770.2
Ajwa	61.5	29125.0	20.0	32000.0	27.0	41000.0	-	-	108.5	32611.6
Others	30.4	8553.9	23.0	15500.0	26.9	8838.2	1.5	2442.0	81.8	10487.2
Total	216.1	14681.2	114.2	17123.7	178.3	15539.0	102.8	3277.7	611.4	13471.7

Source: Study Sample, 2012.

Table 5. Retailers Date Procurement Sources and Prices for the Selected Date Varieties in Saudi Arabia

Date Varity	Owned Farm		Other Farms		Market (auction)		Contracting		All Sources	
	Ton	SR/ton	Ton	SR/ton	Ton	SR/ton	ton	SR/ton	Ton	SR/ton
Sukkari	4.4	27178.6	21.4	14300.0	18.3	19832.3	14.5	11750.0	58.6	16365.0
Khalas	12.8	12150.8	31.9	10590.9	29.4	14079.9	5.0	8000.0	79.1	11976.2
Segae	3.0	19666.7	40.7	5333.3	10.1	16927.7	1.0	10000.0	54.8	8338.5
Barhi	2.7	13333.3	10.7	11531.3	5.1	11250.0	1.0	17500.0	19.5	12011.9
Ajwa	27.5	29500.0	4.0	70000.0	11.4	53622.2	-	-	42.9	39691.0
Other	23.8	10937.5	10.5	10458.3	18.5	15681.3	1.0	8500.0	53.8	12430.7
Total	74.2	19437.0	119.2	11528.9	92.8	20554.3	22.5	10950.0	308.7	16095.9

Source: Study Sample, 2012.

Table 6. Descriptive Analysis for Main Date Marketing Outputs (Quantity, Price, and Value Added).

Items	No.	Min	Max	AVG	STD
Annual quantity of transaction (ton)	294	1	11000	141.11	677.41
Average purchasing price (SR/ton)	297	1500	90000	18685.77	13183.01
Average Selling price (SR/ton)	297	4100	120000	24131.65	15911.801
Average marketing margin[i] (SR/ton)	297	100	40000	5445.87	5304.93

*Significant at 0,05 significance level**Significant at 0,10 significance level
Source: Study Sample, 2012

among marketing channels (at a higher level of course) as well as purchasing channel.

Geographic Factor

Table (8) showed that date price levels (purchasing price and selling price) were significantly higher for Al-Qassim than Riyadh and Al-hasa. These differences could be explained by date variety variations, Sukkari variety common in Al-Qassim.

Educational Level of DMUs

Higher education of DMU manger was accompanied with significant higher quantity of transaction and higher purchasing prices and selling prices (Table 9).

Concentration in date Marketing

Dealers who were concentrated in date marketing purchase and sell dates at significantly higher prices than others whom were not concentrated in date marketing system. Annual quantity of transaction doesn't show significant differences among concentrated or non concentrated dealers (Table 10).

Type of Other Activity Besides Date Marketing

Date dealers whom practice other commercial have significantly higher quantity of transaction than dealers whom practice agricultural activity. Also, they buy and sell dates at a significantly higher prices (Table 11).

Table 7. One Way Analysis of Variance Showing the Impact of Marketing Channels on Date Marketing Outputs in Saudi Arabia

Items	Marketing Channels	N	Mean	Std. Deviation	F	Sig.	Interpretation
Annual quantity of transaction (ton)	Wholesaling	47	113.6[B]	170.5	20.79	0.000	Wholesaling, retailing, exporting versus all
	Retailing	128	56.0[B]	138.7			
	Wholesaling, retailing	106	101.6[B]	174.2			
	Wholesaling, retailing, exporting	9	1704.4[A]	3517.5			
Average purchasing price (SR/ton) Average Selling price (SR/ton)	Wholesaling	47	11490.4[B]	6122.5	10.19	0.000	Retailing And retailing and wholesaling versus wholesaling and wholesaling, retailing, exporting
	Retailing	129	22136.6[A]	13967.2			
	Wholesaling, retailing	106	18462.3[A]	13292.3			
	Wholesaling, retailing, exporting	9	8888.9[B]	4839.8			
Annual quantity Average Selling price (SR/ton)	Wholesaling	47	16576.6[B]	9204.8	6.33	0.000	Retailing and wholesaling, retailing versus wholesaling and wholesaling, retailing, exporting
	Retailing	129	27458.1[A]	16350.3			
	Wholesaling, retailing	106	24036.8[A]	16559.6			
	Wholesaling, retailing, exporting	9	17100.7[B]	15072.7			
Average marketing margin (SR/ton)	Wholesaling	47	5086.2	5158.1	0.96	0.414	-
	Retailing	129	5321.5	4902.6			
	Wholesaling, retailing	106	5574.5	5265.5			
	Wholesaling, retailing, exporting	9	8277.8	11040.0			

- Means followed with same letter do not differ significantly using Duncan's test.
*Significant at 0,05 significance level**Significant at 0,10 significance level
Source: Study Sample, 2012.

Table 8. One Way Analysis of Variance Showing the Impact of the Considered Study Areas on Date Marketing Outputs in Saudi Arabia

Items	Study areas	No.	Mean	Std. Deviation	F	Sig.	Interpretation
Annual quantity of transaction (ton)	Riyadh	74	132.1	258.2	1.42	0.237	-
	Al-Madinah Al-Munawara	98	148.1	302.3			
	Al-hasa	49	294.7	1571.4			
	Al-Qassim	73	37.8	58.7			
Average purchasing price (SR/ton}	Riyadh	74	15528.7[B]	16849.6	4.73	0.003	Al-Qassim versus Riyadh and Al-hasa
	Al-Madinah Al-Munawara	101	19144.1[AB]	8287.9			
	Al-hasa	49	16193.9[B]	11546.5			

Table 8. Cont.

			N	Mean	Std. Dev.	F	Sig.	
		Al-Qassim	73	22924.7^A	14469.2			
Average Selling price (SR/ton)		Riyadh	74	21655.4^B	22299.6	3.13	0.026	Al-Qassim versus Riyadh and Al-hasa
		Al-Madinah Al-Munawara	101	24573.3^{AB}	9633.3			
		Al-hasa	49	20718.4^B	13185.4			
		Al-Qassim	73	28321.9^A	16111.9			
Average marketing margin (SR/ton)		Riyadh	74	6126.7 (39.5 per cent)	7348.5	0.90	0.441	-
		Al-Madinah Al-Munawara	101	5429.2 (28.4 per cent)	5927.9			
		Al-hasa	49	4524.5 (27.9 per cent)	2590.0			
		Al-Qassim	73	5397.3 (23.5 per cent)	2571.2			

- Means followed with same letter do not differ significantly using Duncan's test.
*Significant at 0,05 significance level**Significant at 0,10 significance level
Source: Study Sample, 2012.

Table 9. One Way Analysis of Variance Showing the Impact of Dealer Educational Level on Date Marketing Outputs in Saudi Arabia

Items	Educational level	N	Mean	Std. Deviation	F	Sig.	
Annual quantity of transaction (ton)	Lower than secondary education	149	99.0^B	158.5	10.25	0.000	Higher education versus lower than secondary education and secondary education
	Secondary education	117	78.9^B	196.0			
	Higher education	24	727.3^A	2257.8			
Average purchasing price (SR/ton)	Lower than secondary education	150	15811.2^B	10134.3	6.53	0.002	Higher education versus lower than secondary education
	Secondary education	119	20970.6^{AB}	14962.7			
	Higher education	24	22187.5^A	15708.5			
Average Selling price (SR/ton)	Lower than secondary education	150	20834.0^B	12348.5	6.10	0.003	Higher education versus lower than secondary education
	Secondary education	119	26668.1^{AB}	18151.9			
	Higher education	24	28979.2^A	19000.1			
Average marketing margin (SR/ton)	Lower than secondary education	150	5022.8	5195.2	1.37	0.255	-
	Secondary education	119	5697.5	5224.3			
	Higher education	24	6791.7	6537.8			

Means followed with same letter do not differ significantly using Duncan's test
Significant at 0,05 significance level *Significant at 0,10 significance level
Source: Study Sample, 2012.

Table 10. One Way Analysis of Variance Showing the Impact of Concentration in Date Marketing on Date Marketing Outputs in Saudi Arabia

Items		N	Mean	Std. Deviation	T	Sig.
Annual quantity of transaction (ton)	Yes	79	119.8	241.8	-0.33	0.744
	No	215	149.0	778.9		
Average purchasing price (SR/ton)	Yes	82	21531.1	14285.5	2.31*	0.021
	No	215	17600.6	12603.8		
Average Selling price (SR/ton)	Yes	82	27178.0	16339.8	2.05*	0.041
	No	215	22969.8	15627.9		
Average marketing margin (SR/ton)	Yes	82	5647.0	4958.3	0.40	0.687
	No	215	5369.2	5440.5		

*Significant at 0,05 significance level**Significant at 0,10 significance level

Table 11. One Way Analysis of Variance Showing the Impact of Type of Activity besides Date Marketing on Date Marketing Outputs in Saudi Arabia

Items	Type	N	Mean	Std. Deviation	T	Sig.
Annual quantity of transaction (ton)	Agricultural	68	95.5	191.0	-2.27*	0.026
	Commercial	11	269.5	427.7		
Average purchasing price (SR/ton}	Agricultural	71	20247.2	12592.6	-2.11*	0.038
	Commercial	11	29818.2	21348.6		
Average Selling price (SR/ton)	Agricultural	71	25922.5	14698.8	-1.99*	0.047
	Commercial	11	35281.8	23807.4		
Average marketing margin (SR/ton)	Agricultural	71	5675.4	5219.9	0.13	0.896
	Commercial	11	5463.6	2892.2		

Significant at 0,05 significance level *Significant at 0,10 significance level

- **Length of Experience in Date Marketing**

Table (12) showed the significant correlation between length of experience in date marketing and average purchasing and selling prices of dates in Saudi Arabia.

- Date Waste (loss)

Waste was one of the important key indicator to improve marketing efficiency in general. Al-Madinah Al-Munawara represented the highest date marketing waste out of date marketing purchase (5.8 per cent). For all regions, the date marketing waste out of date marketing purchase was 4.6 per cent with an average price equivalent to about 1775.6 SR/ton.

CONCLUSION

This paper aims to explore marketing output transaction in Saudi Arabia, including marketing characteristics, volume, prices, marketing channels, and loss analysis . Results of the research paper showed that the average annual activity in date marketing amounted to about 820.5 tons of date as a total transaction for an average dealer. This amount contains all date varieties in the major date production regions including Sukkari, Khalas, Segae, Barhi, Ajwa, and other date varieties. The procurement sources were categorized into four sources: owned date farms, direct purchase from other date farms, buying through auction transaction and contracting. The average annual activity in date marketing for a wholesaler is amounted to about 611.40 tons of date as a total transaction from the different purchasing sources with an average price of about SR 13471.7/ ton. The average annual activity in date marketing for a retailer is amounted to about 308.70 tons of date as a total transaction from the different procurement sources with an average price of about SR 16095.90/ ton.

Average date purchasing prices amounted to about SR 17142/ton. The price varies according to date variety and

Table 12. Correlation between Length of Experience in Date Marketing and Date Marketing Outputs in Saudi Arabia

Items	R. coefficient	Level of Significant
Annual quantity of transaction (ton)	R	0.03
Average purchasing price (SR/ton)	Sig R	0.599 0.24
	Sig	0.000**
Average Selling price (SR/ton)	R	0.14
	Sig	0.018**

*Significant at 0,05 significance level**Significant at 0,10 significance level
Source: Study Sample, 2012.

Table 13. Average Date Waste out of Purchasing and its Prices at Different Study Areas in Saudi Arabia

Study Areas	Unit	Min	Max	Weighted. AVG	St. d	C.V
Riyadh	per cent	1	40	3.9	5.6	135.6
	SR/ton	100	10000	831.5	1499.2	163.6
Al-Madinah Al-Munawara	per cent	1	14	5.8	2.2	62
	SR/ton	100	6000	1338	1002	87.6
Al-hasa	per cent	1	9	4.03	1.97	52.8
	SR/ton	1000	9000	2888.5	2128	64
Al-Qassim	per cent	1	7	2.7	1.9	56.5
	SR/ton	500	8500	3229	2422	63
All regions	per cent	1	40	4.6	3.5	93.5
	SR/ton	100	10000	1775.6	2100.4	105.8

Source: Study Sample, 2012.

also according to procurement sources. Date selling prices varies according to date varieties and marketing channel. Ajwa recorded the highest average selling price (55493.41 SR/ton). Sukkari and Segae come after Ajwa (19028.41, 18923.01 SR/ton respectively) .Khalas and Barhi are the cheapest dates sold by the dealers of the sample (12336.81, 11427.73 SR/ton respectively). the marketing margin varies among the study areas, with a minimum of 4524.5 SR/ton in Al-hasa area (23.5 per cent) to a maximum of about 6126.7 SR/ton in Riyadh area (39.5 per cent).

Retailing prices were always higher than wholesaling prices, However marketing spread and price mark up varies according to date variety. These variations may be explained by differences among varieties in the need for marketing services in retailing stage. It was noticed that specialized wholesalers are lower wholesale price oriented, and specialized retailers were higher retail price oriented. This result indicate that date retailing was becoming more oriented to more marketing services, such as high quality packaging, grading, sorting, storage. Waste was one of the important key inductor to improve marketing efficiency in general. Al-Madinah Al-Munawara represented the highest date marketing waste out of date marketing purchasing (5.8 per cent). For all regions, the date marketing waste out of marketing purchasing was 4.6 per cent.

ACKNOWLEDGMENTS

The project team is highly appreciated King Abdulaziz City for Science and Technology (www.kacst.edu.sa) of Saudi Arabia for its financial support and continues encouragement to carry out this research project. This research project is funded thru Strategic Technology Program – the National Plan for Science, Technology, and Innovation (http://nstip.kacst.edu.sa) grant # 600-32.

REFERENCES

Al-Kahtani, Safar and Mohammed ElFeel (2006). "marketing costs for some vegetables and fruit crops, Saudi Arabia," the magazine Alexandria for the exchange of scientific, University of Alexandria, 27: 131-148.

Al-Kahtani,Safar, M. AlQunabet, S. Ismaiel, and H. Hebaisha (2007). Agricultural Marketing in the Kingdom of Saudi Arabia: Existing Situation, Problems, and

Solution. Projects Funded By King Abdulaziz City for Science and Technology.

Al-Kahtani, Safar, S. Ismail, S. Aleid , and H. Bakri (2011). The Prospects and Possibilities of E-Trade of the Saudi Dates and its Economic Role of Supporting Agricultural Sector. Project No EM-2 , Funded by Date Palm Research Center, King Faisal University.

Alshuaibi, A., 2011. The econometrics of investment in date production in Saudi Arabia. Int. J. Applied Econ. Finance,5: 177-184.

Dickey,D. and Fuller, W.(1979). Distribution of the estimators for auto regressive time series with a unit root. Journal of the American Statistical Association, 47: 427-431.

Engle, R.F. and C.W.J. Granger and C.W.J. Granger. 1987. Co integration and Error Correction Representation, Estimation, and Testing. Econometrica 55: 251-76.

FAO, (2014), www.fao.org.

Freedman, David A. (2009). *Statistical Models: Theory and Practice*. Cambridge University   Press.

John baffes. (1991). Some further evidence on the law of one price: the law of one price still holds. American J. of Agricultural Economics, 73:1264-1273.

MOA, Ministry of Agriculture (2014).Agricultural Statistical Year Book. Department of Studies Planning and Statistics, Agricultural Research and Development Affairs, Ministry of Agriculture, Kingdom of Saudi Arabia.

Tomek, W. G; and Robinson, K. L (1985)."Agricultural Product Prices ".Cornell University Press.2nd Edition.

16

Cost and returns of paddy rice production in Kaduna State of Nigeria

Ben-Chendo G.N[1]*, N. Lawal[2], M.N. Osuji[3], I.I. Osugiri[4], B.O. Ibeagwa[5]

[1*,2,3,4,5] Department of Agricultural Economics, Federal University of Technology Owerri, Imo State, Nigeria.

As a result of increasing population growth and urbanization, there is a high and increasing demand for rice, this necessitates the high attention for its production. This research was conducted to determine the profitability of paddy rice production in Chikun Local Government Area of Kaduna State. Data were collected from 60 randomly selected paddy rice farmers using a well structured questionnaire and analyzed using the descriptive statistics, net income and multiple regression models. The result showed that 97% were male, 88% married and had an average household size of 10 people. All respondents had one form of education and their average farm size was 15ha producing about 3.2tonnes of paddy per hectare. Paddy rice production in the area was estimated to have a profit \$902.51 (₦179,600) and a net returns of \$766.83 (₦152,600). Farm size, system of rice cultivation and household size accounted for 78% of the observe variation in the farmer's income. The study however concluded that paddy rice production in the study area is a profitable enterprise and it also recommended that consistent government policies that would favour increase in paddy production, market information, extension service delivery, input subsidization and credit facilities be implemented.

Keywords: Paddy rice, production, profitability, costs, returns, Kaduna

INTRODUCTION

Rice (*Oryza sativa* L.) being the second largest consumed cereal (after wheat) shapes the lives of millions of people. More than half the world's population depends on rice for about 80% of its food calorie requirements. It has become a staple food in Nigeria such that every household; both the rich and the poor consume a great quantity (Godwin, 2012). A combination of various factors seems to have triggered the structural increase in rice consumption over the years with consumption broadening across all socio-economic classes, including the poor. Rising demand is as a result of increasing population growth and income level (GAIN, 2012) coupled with the ease of its preparation and storage. Rice has changed from being a luxury food to a necessity because of its availability and affordability, so consumption will continue to increase with per capita GDP growth, thus implying that its importance in the Nigerian diet as a major food item will increase as economic growth increases (Ojogho and Alufohai, 2010).

Despite the relative importance of rice as Nigerian major food and industrial material, the domestic supply is still considered insufficient to match the consumption demand which leads to huge import to meet local demand. The local production falls short of the demand (Basorum and Fasakin, 2012) hence, leading to augmentation of shortfall through import. According to Ekeleme *et al.*, (2008), and (USAID, 2013), Nigeria consumes 5.4 million metric tonnes of rice annually, of this value, annual domestic output of rice still hovers around 3.0 million metric tonnes leaving the huge gap of about 2 million metric tons to importation.

---

*Corresponding author: Dr (Mrs) Ben-Chendo Glory Nkiruka, Department of Agricultural Economics, Federal University of Technology Owerri, Imo State, Nigeria.
Email: gbenchendo@gmail.com,

This is consistent with the Daramola (2005) that rice importation in Nigeria is around $2.2 Billion which is to the detriment of the scarce foreign exchange reserve. This is an unfortunate scenario for a country that is presumed to be the largest producer of rice in West Africa also depending on imports from countries like Thailand, India and USA etc.

Rice is cultivated in virtually all agro-ecological zones of the country as it constitutes one major cereal crop produced by Nigerian farmers. It covers both the upland and the swamps, depending on the variety (KNARDA, 2007). Traditionally, domestic paddy rice production was limited to flooded system until irrigated rice production was introduced with the development of pump irrigation schemes beginning in the mid-1990s; and this has permitted rice area and production to expand at par with population growth in recent years (West Africa Rice Development Association WARDA, 1997).

Given the crucial role of rice in the food security of urban and rural households alike, development of rice growing has long been considered a priority in Nigeria. The country has adopted a range of instruments designed to protect and increase local production. The Nigerian National Rice Development Strategy (NRDS) set up in 2009 aims to make the country self-sufficient in rice by raising production of paddy rice from 3.4 million tonnes in 2007 to 12.8 million tonnes in 2018. The NRDS outlines three priority areas of focus to achieve this level of production such as improving post-harvest processing and treatment; developing irrigation and extending cultivated lands and making seed, fertilizer and farming equipment more readily available.

In a bid to also achieve rice self sufficiency in line with the rice transformation plan, the Ministry of Agriculture and Rural Development have rolled out a special intervention programme on dry season paddy production plan in 2013. The dry season paddy production is scheduled to take place across ten states of the Federation namely; Kaduna, Kebbi, Zamfara, Kano, Jigawa, Sokoto, Katsina, Bauchi, Gombe and Kogi states. Considering the recent policies and programmes designed by the government to increase paddy rice production in the country, the inadequate supply of paddy for processing by integrated mills, this research work is designed to assess the production of the crop with regard to its profitability in the study area to isolate factors affecting the farmer's income (producers), and invariably the growth of the Nigerian economy.

METHODOLOGY

The study was conducted in Chikun Local Government Area of Kaduna State. Kaduna state lies between latitudes 10^0 21' and 10^0 33' North of the equator and longitudes 7^0 45' and 7^0 75' East of the Greenwich meridian and has 23 local government areas. The state experiences both wet and dry seasons with the wet season commencing in the month of April in the southern part of the state and between May and June in the northern part of the state. The dry season sets in immediately after the rainy season and is characterized by Harmattan (dry and dusty West African trade wind that blows between the end of November and the middle of March) period with a temperature ranging from 18^0C to 26^0C and the heat period with a temperature that ranges from 32^0C to 39^0C.

Chikun Local Government Area shares a common boundary in the North with Igabi and Kaduna North Local Government in the North West with Birnin Gwari and South West with Niger State, Kajuru and Kachia Local Government Area respectively. According to Shaidu, (2008), the climate of the state favours the production of crops such as rice, maize, beans, guinea corn, millet, cotton, yam, carrot, sugarcane, tomatoes, pepper, onions, garden egg plant, lettuce, *Amaranthus* and tobacco. The state is also known for rearing of livestock such as poultry, sheep, goat, cattle and pig.

Chikun Local Government Area was selected purposively on the basis of being a prominent rice producing area in the State. A three step sampling procedure was adopted in the choice of sample for this study. The first step involved the purposive selection of two communities where rice is produced in relatively large quantities, these are; Kujama and Kakau. The second stage was to identify the registered paddy rice farmers with farm sizes of 1ha and above in the two rice producing communities, already selected with the help of Agricultural Development Programme (ADP) extension agents and other rice farmer groups (RIFAN). This list served as the sampling frame for the study. The third stage involved a random sampling of thirty (30) rice farmers from each of the two rice communities bringing the sample size for the study to sixty farmers.

To determine the cost and returns to paddy rice production in Chikun Local government Area of Kaduna State the gross margin model was employed. The gross margin (GM) is the difference between the total revenue (TR) and the total variable cost (TVC). Meanwhile total revenue is the product of paddy rice quantity per unit-bag (Q) and the price of paddy rice per unit-bag (P). The total cost is given by sum of the total fixed cost depreciated (TFC) and the total variable cost (TVC). Mathematically:

$$GM = GR - TVC \text{--eqn 1}$$

Where, GM = Gross Margin (₦/ha)
   GR = Gross Returns (₦/ha)
   TVC = Total Variable Costs (₦/ha)
While the net income model states;

P= Gross Margin – TFC(depreciated)---------------- eqn 2
Where, **P**= Profit or Net income (₦/ha)
TFC = Total Fixed Cost (₦/ha).

And to ascertain the profitability of this venture, the benefit cost ratio was used as stated;

Benefit cost Ratio = $\frac{\text{Total Benefit}}{\text{Total Cost}}$ -------------------- eqn 3

Multiple Regression Analysis

Multiple regression analysis was used to quantitatively determine the effect of postulated independent variables on the dependent variable. The implicit form of the model is presented thus as:

$Y = f (X_1, X_2, X_3, X_4 + ...+..X_n + m)$--------------------eqn 4

Where,
Y = Income (Naira)
X_1 = Farm Size (ha)
X_2 = System of rice cultivation (Dummy: 1 if upland and 0 if otherwise)
X_3 = Farming experience (years)
X_4 = Age (years)
X_5 = Household size (Number of persons)
X_6 = Method of land acquisition (Dummy: 1 if farmer inherited the land and 0 if otherwise)
m = Error Term

The relationship between the endogenous and each of the exogenous variables were examined using four functional forms; Linear, semi log, Exponential and double log while the best fit functional form was selected based on a priori expectations, level of significance of the variables and the coefficients of multiple determination (R^2).

Linear $Y = \beta_0 + \beta_1 X_1 + \beta_2 X_2 + \beta_3 X_3 + \beta_4 X_4 + \beta_5 X_5 + \mu$

Semi log $\text{Log } Y = \beta_0 + \beta_1 \log X_1 + \beta_2 \log X_2 + \beta_3 \log X_3 + \beta_4 \log X_4 + \beta_5 \log X_5 + \mu$

Double log $\text{Log } y = \beta_0 + \beta_1 \log X_1 + \beta_2 \log X_2 + \beta_3 \log X_3 + \beta_4 \log X_4 + \beta_5 \log X_5 + \mu$

Exponential form $\text{Log } y = \beta_0 + \beta_1 X_1 + \beta_2 X_2 + \beta_3 X_3 + \beta_4 X_4 + \beta_5 X_5 + \mu$

RESULTS AND DISCUSSION

Table 1 revealed that majority of the respondent fell within the age bracket of 40 – 49years accounting for 36.72% of the total farming population. This was closely followed by 26.7% of the respondent belonging to the age bracket of 50 – 59years. The mean age of the respondents was 49years and only 6.7% fell below 30years of age. This result implies that majority of the

total farming population in the study area are still within the productive age and can adequately manage and carry out production activities at an optimal level. The table also revealed that 97% of the total paddy rice farmer's populations were male while only 3% were females. This phenomenon could be explained based on the cultural values and religion prevalent in the study area that restricts women to house chores and does not allow them engage in labour intensive farming activities such as rice production. From the study conducted, virtually all the respondents were married, with a total of 88% constituting 53 farmers while only 7 farmers made up the remaining 12% that were single. Farmers within the study area had varying household sizes ranging from 1 to 30 persons. The average family size is 10persons which are relatively small compared to the labour intensive farming activity prevalent in the area. This therefore accounted for the use of hired labour in place of the family labour. Majority of the respondents were not privileged to attain higher education. Respondents who attained secondary school were 33%   the least respondent of 13% were privileged to attain tertiary education the study also revealed that majority of the farmers 54% had primary education and an average of 9years farming experience was recorded. Their farm sizes ranged from 1 to 40 hectares and above and the study revealed an estimation of 15hectares as the mean farm size. Three basic production systems were identified as follows in the study area; the upland, lowland and developed irrigated perimeter with varying constraints and opportunity in terms of water availability, cost of production, variety and quality of crop. The table showed that in the study area 75%, 22% and 3% of the farmers operate in the upland, lowland "Fadama" and irrigated perimeter production systems respectively.

The cost and returns analysis of paddy rice production per hectare of farmland in Chikun Local Government Area of Kaduna state is contained in Table 2. For the purpose of this study, the gross margin analysis and other profitability ratios were used to determine the profitability of paddy rice production on 1 hectare farmland in the study area. This is estimated by adding up the gross revenue less total variable cost. However, the study survey showed that total gross revenue of $1768.8 (₦ 352,000) is generated from sales of paddy rice per hectare. The average cost incurred purchasing seeds, agro-chemicals; fertilizer and packaging are of $126.6 (₦25,200), $183.9 (₦36,600), $229.2 (₦45,600) and $12.1 (₦2,400) respectively. Other variable cost such as planting, fertilizer application, harvesting among others brought the total variable cost to a sum of $866.3 (₦172,400). The gross margin was however estimated to be $902.5 (₦179,600). The net farm income was further estimated by subtracting the total cost (i.e total variable cost + total fixed cost,= ₦199,400 {$1002}) from the gross revenue. This amounted to $766.8 (₦152,600). The

Table 1. Distribution of farmers by socio-economic characteristics

Socio-economic characteristics	Farmers N = 60		
	Frequency	Percentage (%)	Mean
Gender			
Male	58	97	
Female	2	3	
Martital status			
Married	53	88	
Single	7	12	
Age			
20 -29	4	6.7	
30 – 39	10	16.7	
40 – 49	22	36.7	49
50 – 59	16	26.7	
60 years and above	8	13.3	
Household size			
1 - 10 persons	31	52	
11 - 20 persons	26	43	10
21 - 30 persons	2	3	
31 persons and above	1	2	
Educational Background			
No formal education			
Primary education	32	54	
Secondary education	20	33	
Tertiary education	8	13	
Farm sizes			
1 - 10	30	50	
11 – 20	15	25	
21 – 30	8	13	15
31 – 40	4	7	
41 hectares and above	3	5	
Methods of land acquisition			
Inheritance	45	75	
Purchase	16	26.7	
Lease	13	21.7	
Gift	9	15	
Communal ownership	22	36.7	
Farming Experience			
1 - 5	5	8	
6 - 10	30	50	
11 – 15	18	30	9
16 – 20	6	10	
21 years and above	1	2	
Systems of paddy rice cultivation			
Upland	45	75	
Lowland	13	22	
Irrigated	2	3	

Source: Field survey data, 2014.

benefit cost ratio amounted to 1.77 which implies for every ₦1($0.005) in costs the farmer can expect a benefit of ₦1.77($0.008) while the gross margin ratio was estimated to be 0.51. This result implies that for every ₦1($0.005) generated in sales of paddy the, farmer has ₦51($0.256) left over to cover basic operating costs and profit. This indicates that paddy rice production in the study area is profitable. The study therefore, concluded that paddy rice production in Chikun Local Government Area of Kaduna State is a profitable enterprise. There is no alternative to rice in this Local Government area as it is a staple crop and it's processed to another meal called "tuwo chinkafa". However, maize and soya bean are also cereals produced in the state and are not more profitable. According to Sadig et al., (2013) maize production per hectare is highly profitable with 150% profit with a gross ratio of 0.39. This conforms to the study of (Ogaji Abu (2010) Federal University of Technology, Minna,

Table 2. Cost And Returns Analysis For Paddy Rice Production In Kaduna State

Items	Units	Average Per Ha		Value (₦)
		Quantity	Units Price/ Cost (₦)	
REVENUE:				
Paddy rice yield (output)	Kg	3200	110	352,000
TOTAL REVENUE (A)	(₦)			**352,000**
VARIABLE COSTS (INPUTS) :				
Rice seeds	Kg	70	360	25,200
Fertilizer	Bags (50Kg /Bag)	8	5700	45,600
Agro-chemicals (pre and post emergence)	L	14	2400	33,600
Bags	No	60	40	2,400
LABOUR COST:				
Ploughing, harrowing & leveling	MD	3	3000	9,000
Planting	MD	12	300	3,600
Fertilizer application	MD	10	1000	10,000
Weeding Herbicide application	MD	2	1500	3,000
Harvesting / threshing	MD	4	10000	40,000
TOTAL VARIABLE COST (B)				**172,400**
FIXED COST (DEPRECIATION)				
Rent on Land				6,480
Interest on loan				8,000
Depreciation on implement/machines used				12,520
TOTAL DEPRECIATION (C)				**27,000**
TOTAL COST (D)				**199,400**
GROSS MARGIN (A - B) = E				**179,600**
NET RETURNS (E - C) OR (A - D) = F				**152,600**
BENEFIT/COST RATIO (A / D)				**1.77**
GROSS MARGIN RATIO				**0.51**

Source: Field survey data, 2014. ($1 = ₦199).

individual contribution) stated that the lower the gross and operating ratios, the higher the profitability of the farm enterprise and vice versa.

The Effect of Socioeconomic Characteristics on the Income of Paddy Rice Farmers in Chikun L.G.A of Kaduna State

Based on the regression result in table **3**, of the four functional forms estimated, the coefficients in the linear regression were the most significant as it was observed from their t-values, as well as the appropriateness of their signs with relation to *aprori* expectations and the coefficient of multiple determination (R^2). The R^2 for the linear model (0.78) indicated that the explanatory variables investigated highly accounted about 78% of variations or changes in the dependent variable. Based on the afore-mentioned, the linear model was adopted as the lead equation. The linear model is therefore given as:

$Y = -0.923 + 0.182X_1^{**} + 0.00163X_2^{**} + 0.00098X_3^{**} - 0.00134X_4^{**} + 0.598X_5^{**} + 0.583X_6$

The result showed that farm size (X_1), system of rice cultivation(X_2), Farming experience (X_3), household size (X_5) were statistically significant at 5% probability level ($P < 0.05$). It further showed that the method of land acquisition (X_6) was not statistically significant at 5% level of significance.

The significant variables accounted for 78% of the observed variation in the farmer's income. From the regression result, (X_1) which is the farm size was positively related to farmer's income by a constant magnitude of 0.182 for every 1unit increase in the farm size (1m²). This implies that an expansion in the area of farm land increases the level of output made in the course of production, hence the income. The system of rice cultivation used (X_2), farming experience(X_3) and household size (X_5) also gave a positive relationship indicating that with a unit increase in X_2, X_3, and X_5, the income (Y) will increase by a constant amount of 1.6×10^{-3}, 9.8×10^{-4} and 0.598 respectively. This also implies that improving on system of rice cultivation; increase in farming experience increases the profitability. Variable X_4 (age) was significant but negatively related to the level of

Table 3. Multiple Regression Result of the Relationship between Income and Socioeconomic Characteristics

Functions	Constant	Farm Size (X₁)	System of rice cultivation (X₂)	Farming experience (X₃)	Age (X₄)	Household size (X₅)	Method of land acquisition (X₆)	R²	N	F-Value
Linear	-0.923	0.182	0.00163	9.8E- 4	-0.00134	0.598	0.583			
SE	(0.931)	(0.035)	(0.002)	(0.004)	(4.6E - 4)	(0.557)	(0.834)	0.78	60	30.9**
t-value	-0.990	5.179**	0.698**	2.037**	-2.899**	1.073*	0.699			
Semi-log	-8.663	3.243	-0.133	0.777	-0.185	0.117	0.612			
SE	(2.916)	(0.589)	(0.373)	(0.597)	(0.229)	(0.65)	(0.886)	0.74	60	25.6**
t-value	-2.971**	5.509**	-0.356	1.302	-0.803	0.273	0.691			
Double log	-4.152	0.867	0.054	0.233	-0.011	0.265	0.968			
SE	(1.413)	(0.285)	(0.181)	(0.289)	0.111	(0.315)	(0.431)	0.58	60	12.6**
t-value Exponential	-2.939**	3.040**	0.296	0.804	-0.098	0.842	2.253**			
	-1.572	0.031	0.012	3.6E -4	2.2E -4	0.543	1.138			
SE	(0.505)	(0.191)	(1.2E - 2)	(2.6E -4)	(2.5E -4)	(0.303)	(0.453)	0.55	60	10.9**
t-value	-3.108**	1.614	0.958	1.403	-0.892	1.792**	2.514**			

** F-value significant at 5%, ** t – ratio Significant at 5%, figure in ( ) = Standard error (SE)
Source: Field survey data, 2014.

profit made. This implies that a unit increase in age decreases the level of profit made by a constant magnitude of 0.00134. This is contrary to a priori expectations and can be attributed to the inability of elderly people to adopt new technologies to ensure higher output.

Variable X₆ (method of land acquisition), positively related to the farmer's income was however not statistically significant. This implies that the method of land acquisition did not matter but the level of farming experience could influence the farmer's income.

The analysis finally indicates that in the absence of all the independent variables investigated, the farmer's net income will decrease by a constant magnitude of 0.923.

CONCLUSION

Rice production has become a major source of livelihood for farmers in Kaduna state not only providing them with basic food requirement but also generating income for farmers through the sales of paddy rice, increasing the number of jobs created particularly at the rural communities and contributing to the growth of the economy by increasing the Gross Domestic Product (GDP) of the country. Paddy rice production in Kaduna state has not reached it maximum however, the major findings of this study showed that Kaduna State has great potentials for rice production. At all levels of operation, the study revealed that paddy rice production in the study area holds a promising prospect for investors as evident

in the net returns obtained, the gross margin ratio and benefit cost ratios. All these profitability ratios estimated proved positive and hence depict good profit element for paddy rice farmers in the area. This work therefore, recommends that, there should be consistent policy support by government to transform the rice farmers' mindset from seeing rice as only a food to a cash crop through the provision of farmer education on rice cultivation systems, extension service delivery, credit facilities available and the minimization of risk associated with high level of price fluctuation especially during bumper harvests.

REFERENCES

Basorun, JO, Fasakin, JO (2012) Factors Influencing rice Production in Igbemo-Ekiti region of Nigeria. *J. of Agric., Food and Env'tal Sc.* Vol. (1).

Daramola B (2005).Government Policies and Competitiveness of Nigerian Rice Economy". Paper presented at the workshop on *Rice policy and food security in sub-Saharan Africa*, organized by WARDA, Cotonou, Republic of Benin. November, 07-09.

Ekeleme F, kamara AY, Omoigui LO, Tegbaru A, Mshelia J. Onyibe JE (2008) Guide to rice production in Borno state Nigeria" IITA/Canadian Inter. Dev. Agency (CIDA), Ibadan. Vol 1. pp 2 - 14.

Global Agricultural Information Network. *Nig. Grain and Feed Ann. Report.* GAIN Report Number: NI1204, U.S. Consulate, Lagos; 2012.

Godwin U (2012). *Rice farm &milling plant:* Sure money spinner. Available at: http://nationalmirroronline.net/new/rice-farm-milling-plant-sure-money-spinner/.

Kano State Agricultural and Rural Development Authority (KNARDA). *The planning in upgrading of Rice Production in Kano State.* A package prepared by Marditech Corporation Sdn. Bhd. Malaysia for KNARDA. pp 1- 44; 2007

National Rice Development Strategy (NRDS). A working Document prepared for *"The Coalition for African Rice Development"* on the fourth Tokyo International Conf. on African Dev't (TICAD IV), Yokohama, Japan. pp 6 – 11; 2009

Ojogho, O. Alufohai, G O (2010) *Impact of Price and Total Expenditure on Food Demand in South-Western Nigeria. Afri. J. of Food, Agric., Nutr. and Dev't.* Vol. 10, No. 11, pp. 4350 – 4363. ISSN: 1684-5358.

Sadiq, MS., Yakassi MT, Ahmad MM, Lapkene TY, Abubakar M (2013) Profitability and Production Efficiency of Small-scale Maize Production in Niger State, Nigeria. *J. of Appl. Phys.* (IOSR-JAP) Vol. 3, Issue 4. Pp 19-23

United States Agency for International Development (USAID). Global Food Security Response. West Africa Rice Value Chain Analysis, *"Global Food Security Response Nigeria Rice Study"*. Available online at www.pdf.usaid.gov/pdf_docs/pnaea873pdf; 2013

West African Rice Development Association. "Rice Production, Marketing and Policy in Nigeria". Occasional Paper WARDA, Abidjan Cote D'Ivoire, 3(2): pp.14-16; 1997

17

Willingness to pay for native pollination of blueberries: A conjoint analysis

Thomas Stevens[1], Aaron K. Hoshide[2], Francis A. Drummond[3,4]

[1*] Department of Resource Economics, 224 Stockbridge Hall, University of Massachusetts, Amherst, MA 01003, USA
[2] School of Economics, 206 Winslow Hall, University of Maine, Orono, ME 04469, USA
[3] School of Biology and Ecology, 305 Deering Hall, University of Maine, Orono, ME 04469, USA
[4] University of Maine Cooperative Extension, 5722 Deering Hall, University of Maine, Orono, ME 04469, USA

This study estimates blueberry consumer reaction to a potential honey bee Colony Collapse Disorder (CCD) management strategy; increased reliance upon native pollinators like the common Eastern Bumble bee (*Bombus impatiens*). A survey of 498 consumers was conducted using Amazon's Mechanical Turk. Respondents were asked to rate on a scale of 1 to 5, four different blueberry "packages" each containing five attributes; price, pollination method (native bee, commercial honey bee), fresh or frozen, produced in or out of state and variety (wild, cultivated). Statistical analysis suggests that the average consumer surveyed was willing to pay between $0.51 and $0.74 extra per dry liter for native pollination. Consumer willingness to pay of $0.51 extra per dry liter for an average hectare of blueberries was conservatively 1.75 times the annual cost per hectare for producers to plant wildflower pastures for native bees. Consequently, native pollination may be an economically viable alternative for blueberry producers facing the consequences of CCD and other causes of increased honey bee colony losses.

Keywords: Colony Collapse Disorder, CCD, Conjoint Economic Analysis, Native Bee Pollination, Willingness to Pay, Blueberry, Bee Pasture Costs

INTRODUCTION

Blueberries are often pollinated by commercial honey bees (Isaacs and Kirk, 2010; Rose et al., 2013; Hoshide et al., in prep). This is despite blueberries (the various species of commercially grown and sold as "blueberries") being a native crop with highly co-evolved bee pollinator species being abundant in most of these unique agroecosystems (Jones et al., 2014). Because of honey bee Colony Collapse Disorder (CCD), the commercial honey bee colony population has been recently declining and suffering heavy annual losses (vanEngelsdorp et al., 2009; Ratnieks and Carreck, 2010). For example, in 2012 CCD appears to have been associated with the loss of about 30 to 50 percent of U.S. commercial honey bee colonies. Since bee pollination currently accounts for significant U.S. crop value, CCD and other sources of honey bee colony losses could produce very serious economic problems resulting in reduced food supply and higher food prices (USDA/ARS, 2012).

---

*Corresponding author: Dr. Thomas Stevens, Department of Resource Economics, 224 Stockbridge Hall, University of Massachusetts, Amherst, MA 01003, USA. E-mail address: tstevens@resecon.umass.edu

The objective of this paper is to estimate blueberry consumer reaction to one potential pollination strategy in light of uncertainty of availability of honey bees; increased reliance upon native bee pollinators like andrenids, halictids, megachilids, and bumblebees (Jones et al., 2014, Bushmann and Drummond, in press). In particular, we use conjoint analysis (Novotorova, 2007) to examine the tradeoffs consumers are willing to make between blueberries having five different attributes: pollination method (native or commercial), price, frozen or fresh, produced in or out of state, and variety (wild or cultivated). We then compare such consumer support for native bees to estimated costs of establishing native bee pastures near blueberry fields.

We begin with a brief discussion of the relationship between honey bee colony losses, native pollination and blueberry production. The conjoint analysis of consumer tradeoffs is then conducted with the result that the average consumer surveyed was willing to pay between $0.51 and $0.74 extra per dry liter for native bee pollination. We then estimate the amount of consumer willingness to pay (WTP) per crop area to determine the amount of support for encouraging native bees on-farm assuming such support is transferred from blueberry consumers to producers.

Background

Both wild (native) bees and commercial honey bees have declined globally over the past several decades (NRC, 2007), while demand and plantings of pollinator-dependent crops have increased (Aizen & Harder, 2009). While native bees are overwhelmingly responsible for global crop pollination (Garibaldi et al., 2013), commercial managed honey bees are important pollinators especially for larger, continuous production areas found in crops such as wild (lowbush) blueberry (Asare, 2013) and cultivated (highbush) blueberry (Isaacs and Kirk, 2010). Native bee declines have been attributed to loss of pollen host plants (Scheper et al., 2014). Commercial honey bees are not native to the U.S. but they pollinate about one-third of all crop species grown in the U.S. Although extensive colony losses have been common throughout history, since 2006 adult bees began to abandon seemingly healthy hives and colony losses in the U.S. started exceeding usual losses (vanEngelsdorp et al., 2010; USDA/ARS, 2012).

There are several theories about the cause of CCD. According to the first annual report of the U.S. Colony Collapse Disorder Steering Committee, published in July 2009 (as cited in Ratnieks and Carreck, 2010), the recent declines are "unlikely to be caused by a single unknown pathogen" (p. 153). It is more likely that the declines are caused by a variety of factors in combination, , but all of the multiple stressors appear to be exacerbated by the parasitic *Varroa* mite (Neumann and Carreck, 2010). Abiotic factors are also potential stressors. One possible

factor is unfavorable weather patterns that disrupt foraging or stress colonies (Drummond et al., 2012). And, there have been recent studies, mainly in Europe, that have linked newer systemic insecticides with declining bee health (Goulson, 2013; Goulson et al., 2015). Several European countries have suspended the use of certain pesticides believed to be responsible for acute poisoning of honey bees; though, according to the USEPA (2012), the cases of acute poisoning leading to these bans have not been linked to Colony Collapse Disorder. Unfortunately there is no known cure for CCD and concern about the potential economic impact of CCD and other causes of colony losses continues to increase.

Most of the literature on the economic value of pollination uses crop supply and demand analysis. For example, Southwick and Southwick (1992) used econometric demand and supply models to estimate the gains to consumers in the form of lower prices for agricultural products as a result of increased yields due to pollination by honey bees. Their estimate of the annual value of honey bees to U.S. consumers range between $1.6 billion dollars (in 1986 dollars) assuming that there are replacement pollinators and $8.3 billion dollars (in 1986 dollars) assuming there are no replacement pollinators. However, a crucial assumption in this study is a perfectly elastic (horizontal) long-run aggregate supply curve for agricultural commodities. This implies that farmers and resources will be shifted toward or away from a particular crop in response to price change, without additional costs. Ultimately price will be stabilized by market entry/exit and shifting of resources. But the assumption that price is equal to supply eliminates producer surplus.

Gallai et al. (2009) use a bio-economic approach that focuses on two factors for 100 crops to estimate the economic consequences of pollinator decline. One factor is crop dependence on pollinators. This is simply the proportion of the total crop that is reliant on pollination; it can be thought of as the percentage difference in crop yields with and without pollinators. The other factor is the producer's ability to adjust production in the face of pollinator declines. When focusing solely on the impacts on agriculture, Gallai et al. (2009) estimate the total economic "Production Value" of pollination to be €153 billion, making up 9.5% of the total value of the world agricultural production used for human food consumption. This value only considers the relationship between pollination and crop production and assumes a constant price. When changes in price are considered, Gallai et al. (2009) estimate the consumer surplus due to pollination to be between €190 and €310 billion.

Winfree et al. (2011) use an "Attributable Net Income" method to value pollination services. As they state, "the attributable net income method improves upon previous methods in three ways: (1) it subtracts the cost of inputs to crop production from the value of pollination, thereby not attributing the value of these inputs to pollination; (2) it values only the pollination that would be utilized by the crop plant for fruit production, thereby not valuing pollen deposited in excess of the plant's requirements; and (3) it can attribute value separately to different pollinator taxa..." (p. 80). Using a production function estimated from producer surveys, Hoshide et al. (in prep) found marginal Attributable Net Income of pollination diminished from $309 to $156/ha for every 2 rented honey bee hives added per hectare up to 10 hives per hectare for Maine wild blueberries.

On the other hand, Allsopp et al. (2008), use "Replacement Cost" (RC) to estimate the value of pollination as an ecosystem service and find the value of pollination services to be significantly higher than market prices for commercial pollination. The RC method values pollination by estimating the cost of an alternative means of pollination. Realistic replacement methods for honey bee pollination involve either substitution with managed non-honey bee pollinators such as industrially reared bumblebees or pollination by mechanical means (Allsopp et al, 2008). Land surrounding crops can also be managed to increase native bee abundance to enhance fruit set (Blaauw and Isaacs, 2014) by clearing land to create early successional habitat (Nicholls and Altieri, 2013; Garibaldi et al., 2014), actively planting bee floral pastures (Carreck and Williams, 2002; Haaland et al., 2011), and by providing nesting materials (Wratten et al., 2012).

Instead of using previously mentioned econometric crop demand and supply models, this study is unique in that we use a stated preference approach to measure consumer willingness-to-pay, WTP, for native as opposed to commercial pollination for both wild (lowbush) and cultivated (highbush) blueberries. The stated preference method is particularly well suited for this study because it can measure the value of goods and services not yet in the marketplace. Moreover, the stated preference approach has often been used to value attributes of agricultural commodities. Darby et al. (2006), for example, conducted a discrete choice experiment to value strawberry characteristics. The variables used in their choice experiment included price, location of production, size/type of producing firm, and a product freshness guarantee. Olesen et al. (2010) conducted a choice experiment to elicit Norwegian consumers' willingness to pay for organic and animal welfare labeled salmon. Loureiro et al. (2001) also conducted a choice experiment to assess consumer choice of eco-labeled, organic, and regular apples. Another study that assessed the importance of apple attributes to consumers was conducted by Novotorova (2007) who examined consumer preference for genetically modified apples.

There have been many other stated preference studies of food attributes but only two previous studies have examined consumer willingness-to-pay for blueberry attributes. Hu et al. (2009, 2011) used an in-store conjoint survey to examine six blueberry products.

For each, consumers were asked to choose among four product attributes: organic or not, produced within state or not, sugar free or not, and price. Their results indicated that "local products and organic formulations generally received positive willingness to pay across all products" (p.47). Shi et al. (2011) surveyed 772 individuals about their WTP for blueberries with four attributes: freshness (fresh, frozen), organic or conventional production, local production and price ($2.72 to $10.88/dry liter). Their results showed a consumer preference for locally produced berries, but "less than 50 percent of respondents demonstrated positive premiums for organic blueberries" (p.2). We did not find any published research that examined consumer WTP for pollination method—this study attempts to fill that gap.

METHODS

This study is based on the premise that although there is some uncertainty, native wild bees and other insects can provide adequate pollination of blueberries (Isaacs and Kirk, 2010; Asare, 2013; Bushmann and Drummond, in press) However, increased wild bee habitat management such as planting flowers, minimum mowing of meadows, and less use of pesticides is required for long-term reliance upon native pollinators for sustainable production of blueberries (Rose et al., 2013; Hanes et al., 2013). Consequently, increased use of native pollination may result in higher production costs and higher market prices.

Yet, there is virtually no information about the tradeoffs blueberry consumers are willing to make, if any, between pollination method and product price. To examine these tradeoffs we employed a conjoint stated preference method that asked consumers to rate several blueberry "packages" each with different attributes. The traditional conjoint model is based on the following theoretical relationship between consumer utility, ratings and blueberry attributes:

$$U_i = Y_i = V(X^K) + P_x = b_0 P_x + b_1 X_1^1 + \ldots\ldots b_n X_n^1 + e_i \qquad (1)$$

Where U_i is individual i's utility for an attribute bundle; Y_i is the individual's rating, $V(.)$ is the non-stochastic component of the utility function, X^K is a vector of attribute levels, P_x is the price for the attribute bundle or package X and b is the marginal utility or weight associated with each attribute.

Setting the total differential of (1) to the point of indifference and solving:

$$dU_i = b_0 dP_x + b_i dX_1^1 + \ldots\ldots = 0 \qquad (2)$$

Yield marginal rates of substitution for the attributes X.

Since a price attribute, P_x, is included, the marginal utilities of all attributes can be rescaled into dollars, and marginal willingness to pay for each attribute may be derived:

$$dP_x = -b_1 dX_1^1/b_0 \quad \text{or}$$
$$dP_x/dX_1^1 = -b_1/b_0 \qquad (3)$$

The stated preference conjoint survey used in this study was developed using verbal protocol methods. The verbal protocols were conducted in three stages and the survey was revised after each protocol session. An online pilot survey was then undertaken and the final survey of 498 individuals was conducted in 2013 using Amazon's Mechanical Turk. Mechanical Turk is "an online labor market where requesters post jobs and workers choose which jobs to do for pay" (Mason and Suri, 2012, p. 1). Respondents are not necessarily representative of the population, but as noted by Mason and Suri (2012), "there have been a number of studies that validate the behavior of workers as compared to offline behavior"(p.17). And, compared to other survey methods Mechanical Turk provides access to a large, very diverse group of subjects.

Respondents answered questions about their purchase of blueberries, price paid and maximum price per dry volume they would be WTP for both fresh and frozen berries. We also asked about knowledge of CCD, membership in environmental organizations, age, education, state of residence and income. Respondents were asked to rate on a scale of 1-5 four different blueberry purchase options. A rating of 1 indicated options the respondent would definitely not purchase while a 5 indicated options, if any, the respondent would definitely purchase. A copy of the conjoint portion of the survey questionnaire is presented in the Appendix.

Each blueberry option or package contained five attributes; price, pollination method, fresh or frozen, produced in or out of state and variety (wild, cultivated). The number of possible options was reduced to 16 using the standard fractional factorial design. Each respondent received four options and so there were four survey blocks. A short description of pollination methods, CCD, and the difference between wild and cultivated blueberries was presented to each respondent prior to the rating question (see Appendix).

An ordered logit model was used to analyze the survey data. This model is appropriate when there is a naturally ordered preference scale that is represented by a discrete ordered observed outcome, such as the rating scale system used in our survey (Greene, 2012). The ordered logit model is built around the latent regression model. In the latent model, the unobserved continuous dependent variable (y^*) is a function of both observed independent variables ($\mathbf{x}$) and the unobserved disturbance (ε):

$$y^* = \ln(\beta_i X_i + \varepsilon_i) \qquad (4)$$

What we do observe is the ordered variable (y), which in our case is represented by ratings. The unobserved latent dependent variable (y^*) is tied to the observed variable (y) by the observation rule:

$$y = 1 \text{ if } y^* \le \mu_1$$
$$y = 2 \text{ if } \mu_1 < y^* \le \mu_2$$
$$y = 3 \text{ if } \mu_2 < y^* \le \mu_3$$
$$y = 4 \text{ if } \mu_3 < y^* \le \mu_4$$
$$y = 5 \text{ if } \mu_4 < y^* \le \mu_5 \qquad (5)$$

where y is the observed rating and the μ's are unknown threshold parameters to be estimated with β. The thresholds are assumed to be strictly increasing and restricted to be positive. We assume that ε is distributed logistically across observations. The first threshold is normalized to zero resulting in one less parameter needing to be estimated (Baetschmann, 2012).

The probabilities associated with the ratings j=1,2,3,4,5 are found using the cumulative distribution function which is given by:

$$\Pr(y_i \le j | \mathbf{x}_i) = \ = \ = \ = \ \Phi(\mu_i - \textstyle\sum x_i \beta) \qquad (6)$$

To find the probabilities of a specific rating j occurring, the probability from the cumulative distribution for rating j-1 is subtracted from the cumulative distribution probability for rating j:

$$\Pr(y_i = 1 | \mathbf{x}_i) = \Phi(\mu_1 - \textstyle\sum x_i \beta)$$
$$\Pr(y_i = 2 | \mathbf{x}_i) = \Phi(\mu_2 - \textstyle\sum x_i \beta) - \Phi(\mu_1 - \textstyle\sum x_i \beta)$$
$$\Pr(y_i = 3 | \mathbf{x}_i) = \Phi(\mu_3 - \textstyle\sum x_i \beta) - \Phi(\mu_2 - \textstyle\sum x_i \beta)$$
$$\Pr(y_i = 4 | \mathbf{x}_i) = \Phi(\mu_4 - \textstyle\sum x_i \beta) - \Phi(\mu_3 - \textstyle\sum x_i \beta)$$
$$\Pr(y_i = 5 | \mathbf{x}_i) = 1 - \Phi(\mu_4 - \textstyle\sum x_i \beta) \qquad 7)$$

In order to estimate this model a log-likelihood function must be formed which is obtained by summing the probabilities over the sample.

Estimation of the regression coefficients can be found by maximizing the log likelihood function (Greene and Hensher, 2010). Independent variables in this model were blueberry option attributes and selected socio-economic characteristics of respondents defined in Table 1.

Budgets were constructed for both wild and planted native bee floral pasture strips including both variable costs such as fuel and seed as well as fixed costs such as equipment depreciation and land as specified in Kay (2011). The wild pasture strip budget assumed an indefinite stand life accompanied by annual late fall mowing to prevent succession to forest with no direct seeding of flowering plants, instead relying on native, natural establishment of indigenous or exotic flowering plant species. Since natural or planted bee pastures do not generate revenue for the farm like a cash crop, it was assumed that the costs of such bee forage have to be covered by the cash crop or externally subsidized. Economic data used for budgets was collected from cooperating lowbush blueberry producers surveyed at

Table 1. Variables Defined

Price:	price of blueberry package per pint (per dry liter)
	$1.25, $2.75, $4.25, $5.75 ($2.27, $4.99, $7.72, $10.44)
Native:	0 = pollination method of blueberries is native wild bees
	1 = pollination method of blueberries is commercial honey bees
Fresh:	0=frozen blueberries
	1=fresh blueberries
In_state:	0= blueberries produced out of state
	1= blueberries produced in state
Wild variety:	0= cultivated blueberries
	1= wild blueberries
Rating:	1=definitely not purchase blueberry package
	2= probably not purchase blueberry package
	3= may or may not purchase blueberry package
	4= probably purchase blueberry package
	5= definitely purchase blueberry package
Somewhatinf:	0= Respondent was at most minimally informed about the bee problem
	1= Respondent was at least somewhat informed about the bee problem
Household:	Number of people in household
Gender:	0=Respondent female
	1=male
Age:	1= respondent is between 18 and 24 years of age
	2= respondent is between 25 and 34 years of age
	3= respondent is between 35 and 44 years of age
	4= respondent is between 45 and 54 years of age
	5= respondent is between 55 and 64 years of age
	6= respondent is 65 years of age or older
Bachelor:	0 = Respondent's Education level is: some college or less
	1 = Respondent's Education level is bachelor degree or higher
Region:	2 = Respondent resides in the Northeast:
	ME, NH, VT, MA, RI, CT, NY, PA, NJ
	3 = Respondent resides in the Midwest:
	WI, MI, IL, IN, OH, MO, ND, SD, NE, KS, MN, IA
	4 = Respondent resides in the Southeast:
	DE, MD, DC, VA, WV, NC, SC, GA, FL, KT, TN, MS, AL, AR, LA
	5 = Respondent resides in the Southwest:
	OK, TX, AZ, NM
	6 = Respondent resides in the West:
	ID, MT, WY, NV, UT, CO, AK, WA, OR, CA, HI
inc4500:	
	1= Respondent's annual household income is under $45,001
inc45001-85000:	1 = Respondent's annual household income is between $45,001 & $85,000
inc85001:	
	1 = Respondent's annual household income is greater than $85,000

grower meetings and on-farm between 2012 and 2013 (n=80; about 20% of producers conducting management operations). About 55% of surveyed producers leave areas on their farms for natural wildflower establishment, while only 15% actively plant bee pastures for native bees (Hanes et al., 2013).

Planted bee pastures were modeled using tilled methods based on experimental plantings at two research farms in Jonesboro and Orono, Maine, and on-farm at two cooperating wild blueberry producers in Blue Hill, Maine. Planted areas of 0.046 hectares were cleared, tilled, stale-seed bedded, limed, seeded, and roller-packed May to June 2012. Plots of clover mix, wildflower, and control were mowed in 2012 and 2013 to control competition from annual weeds. All material, time, and machinery expenses incurred were recorded in detail during site preparation and the initial establishment phase. Expenses were categorized and averaged across sites for use in enterprise budget modeling. Economics of all planted bee pasture were analyzed for a three-, five-, and ten-year stand life before having to re-establish the planting due to weed competition.

Table 2. Results for Ordered Logit Regression, With Socio-Demographics

Variables	coefficients	Std. Err.	z value	P-value	odds ratio	% change in odds
price	-0.766	0.030	-25.14	0	0.465	-53.5
native	0.310	0.081	3.82	0	1.363	36.3
fresh	1.424	0.093	15.31	0	4.153	315.3
in_state	0.324	0.081	3.99	0	1.382	38.2
wild_variety	0.080	0.081	0.98	0.328	1.083	8.3
somewhatInf	0.226	0.096	2.35	0.019	1.253	25.3
household	0.043	0.032	1.36	0.172	1.044	4.4
gender	0.243	0.085	2.87	0.004	1.275	27.5
age_1	1.183	0.341	3.47	0.001	3.265	226.5
age_2	1.307	0.331	3.95	0	3.696	269.6
age_3	1.378	0.341	4.05	0	3.967	296.7
age_4	1.438	0.349	4.13	0	4.212	321.2
age_5	0.909	0.356	2.55	0.011	2.483	148.3
bachelor	-0.033	0.111	-0.3	0.764	0.967	-3.3
inc45000	-0.102	0.096	-1.06	0.29	0.903	-9.7
inc85001	-0.115	0.122	-0.94	0.348	0.892	-10.8
region_2	-0.093	0.131	-0.71	0.478	0.911	-8.9
region_3	-0.156	0.130	-1.2	0.229	0.855	-14.5
region_4	-0.199	0.122	-1.64	0.102	0.819	-18.1
region_5	0.005	0.166	0.03	0.975	1.005	0.5
/cut1	-2.509	0.370				
/cut2	-1.023	0.366				
/cut3	0.074	0.365				
/cut4	1.774	0.367				

RESULTS AND DISCUSSION

About 53% of the 498 respondents were male, 47% had at least a bachelor degree, the median income was between 45 and 60 thousand dollars and the median age was between 25 and 34 years old. Thirty-nine percent of respondents had heard of the CCD problem before the survey.

Results derived from the ordered logit model are presented in Table 2. All model blueberry attribute coefficients except for the wild variety attribute were statistically significant and had the anticipated sign. Statistically significant socio-economic characteristics include being at least somewhat informed about the CDC problem, males and people under 65 years old. Education, income and the regional dummy variables were not statistically significant factors. Marginal WTP

for changes in each attribute is the ratio of the estimated attribute and price coefficients, $B_{attribute}/B_{price}$. Therefore, the average respondent was WTP \$0.74/dry liter more for native pollination. Respondents were WTP \$3.38/dry liter more for fresh berries and blueberries grown in state commanded a \$0.76/dry liter premium, all else held constant at mean values.

The odds ratios (see Table 2) indicate the probability of choosing a higher rating. For example, as shown in Table 2, the odds of an individual choosing a higher rating is about 36.3% greater for blueberries pollinated by native insects. Similarly, the odds of choosing a higher rating are about 53.5% less for a one dollar increase in price, 315% greater for fresh blueberries and 38% higher for blueberries grown in state.

Hypothetical bias wherein respondents may select more expensive options than they would actually buy is a

Table 3. Results for Binary Logit Regression, With Socio-Demographics

Variables	coefficients	Std. Err.	z value	P-value	odds ratio	% change in odds
price	-0.79421	0.041855	-18.98	0	0.451938	-54.8
native	0.219815	0.105759	2.08	0.038	1.245846	24.6
fresh	1.613654	0.13356	12.08	0	5.021123	402.1
in_state	0.313476	0.105641	2.97	0.003	1.368173	36.8
wild_variety	0.113471	0.109697	1.03	0.301	1.120159	12
somewhatinf	0.171702	0.124687	1.38	0.168	1.187324	18.7
household	-0.00576	0.04145	-0.14	0.889	0.994257	-0.6
gender	0.268885	0.110054	2.44	0.015	1.308504	30.9
age_1	1.199092	0.447543	2.68	0.007	3.317102	231.7
age_2	1.328745	0.434835	3.06	0.002	3.776302	277.6
age_3	1.356722	0.4473	3.03	0.002	3.883442	288.3
age_4	1.29215	0.456141	2.83	0.005	3.640604	264.1
age_5	1.06668	0.464471	2.3	0.022	2.905716	190.6
bachelor	0.082509	0.144909	0.57	0.569	1.086008	8.6
inc45000	-0.2557	0.124636	-2.05	0.04	0.774371	-22.6
inc85001	-0.16015	0.157516	-1.02	0.309	0.852012	-14.8
region_2	-0.24732	0.170308	-1.45	0.146	0.780892	-21.9
region_3	-0.326	0.167	-1.960	0.050	0.721578	-27.8
region_4	-0.364	0.157037	-2.32	0.02	0.694756	-30.5
region_5	-0.07805	0.214515	-0.36	0.716	0.924919	-7.5
_cons	0.208936	0.480076	0.44	0.663	1.232366	

potential problem with stated preference survey data. One of the methods often used to reduce hypothetical bias is certainty calibration where only those who would definitely purchase an option are counted as "yes" responses (Murphy and Stevens, 2004). A sensitivity analysis was conducted by estimating a dichotomous choice logit model with the dependent variable defined as 1 if the individual would "definitely" purchase the option (rating of 5), and 0 otherwise.

The binary results presented in Table 3 are very similar to those in Table 2. The average respondent was WTP $0.51 more per dry liter for native pollination, $0.73 extra for berries grown in-state and $3.71 more for fresh blueberries.

This study contributes to the literature on the valuation of bee pollination; no prior study on the value of bee pollination in the consumer choice framework has been found to be published. Through a conjoint analyses of data obtained from an online survey valuing blueberries with method of pollination as an attribute, a value for pollination was obtained. This study suggests that price, variety, pollination method, and place produced all influence consumer preferences toward blueberries.

Specifically, consumers are willing to pay a positive premium for fresh blueberries, blueberries pollinated by native bees, and blueberries produced in state. Consumers were willing to pay the most for fresh blueberries, over four times the willingness to pay for any other attribute. Consumers were willing to pay a similar amount extra for blueberries pollinated by native bees and for blueberries produced in state, with a slightly higher marginal value for blueberries produced in state.

The positive premium for fresh blueberries and blueberries produced in state is not surprising as multiple past studies have found similar results with most produce. The key result that has not been shown in a consumer choice framework before is that consumers were willing to pay a positive premium for blueberries pollinated by native bees. Values of willingness to pay for native pollination ranged from $0.51 to $0.74 per dry liter depending on the model used.

Model results clearly indicate consumers are willing to pay a significant premium for native pollination (Table 4). For example, the per area value ($5,346/ha) of a conservative $0.51/dry liter premium is 1.75 times the annual cost of native bee wildflower pasture ($3,052/ha)

Table 4. Consumer and Producer Willingness to Pay (WTP) for Native Bee Pollination Compared to Annual Costs of Native Bee Pasture for U.S. Blueberries

Crop / Pollination Option	Consumer WTP ($/dry liter)	Yield (dry liters/ha)	Consumer WTP per harvested blueberry ha	Producer WTP per harvested blueberry ha	Stand Life (yrs)	Annual cost [c] per harvested blueberry ha
Wild (lowbush) &	$0.51	10,513 [a]	$5,346	$140 [b]	-	-
cultivated (high-bush) blueberry	$0.64	-	$7,828	-	-	-
Native bee pasture						
Clover	-	-	-	-	3	$2,768
	-	-	-	-	5	$1,964
	-	-	-	-	10	$1,371
Wildflower	-	-	-	-	3	$3,052
	-	-	-	-	5	$2,137
	-	-	-	-	10	$1,458
Natural [d]	-	-	-	-	Lifetime	$494

[a]Wild and cultivated blueberry average weighted by production area.
[b]Wild blueberry only.
[c]Includes both variable and fixed costs.
[d]Assumes annual mow to sustain early successional wild native bee forage.

assuming a shorter than anticipated three-year stand life. Given wild blueberry producers surveyed in Maine have limited willingness to invest in practices to encourage native bees ($140/ha), external subsidization of such practices (e.g. from consumer premiums) appear necessary to encourage wider adoption in both the wild (Hanes et al., 2013) and cultivated (Blaauw and Isaacs, 2014) blueberry industries.

However the costs of native bee pasture may not be so easily covered by consumers for three reasons. First, actual consumer WTP may be less than elicited in our survey, pricing more consumers out of the market for blueberries. Second, consumer premiums paid for blueberries labeled as native bee friendly may only be applicable early in the adoption of native bee conservation practices. Premium values calculated here assume the entire (wild and cultivated) blueberry industries have adopted native bee floral pastures. In this scenario, it is likely the initial premium would go back toward zero. Finally, a third consideration is that while this premium may be more easily transferred from consumers to producers in the fresh blueberry retail market, almost half of the total blueberry industry is frozen (USDA, NASS, 2010-2013). For both the wild and cultivated blueberry frozen market as well as the fresh wholesale market, there would likely be more administrative costs to facilitate transfer of this premium from consumers to producers. This would reduce the premium left for producers to invest in native bees. Assuming consumers are willing to pay at least a $0.51/dry liter native bee friendly premium for at least half of total blueberry production and if administrative costs are low, native bee pollination may be an economically

viable alternative for blueberry producers facing higher pollination costs due to projected continued high levels of commercial honey bee colony losses. When interpreting our results it is also important to note that although willingness-to-pay has been used in many previous studies, this method can produce biased results. However, the uncertainty adjustment technique employed here generally reduces or eliminates hypothetical bias (Murphy and Stevens, 2004). The Mechanical Turk survey method has been tested by others (see Masori and Suri, 2012), respondents were generally representative of the US population, and our value estimates exceed estimated cost of native pollination by a substantial margin (1.75). Consequently we believe our results are useful for producer decision making. Similar consumer WTP studies should be conducted for other pollinator-dependent crops to see if, like blueberry, native bee conservation premiums adequately cover the costs of such program implementation by producers. Consumer premiums to support alternative pollination practices by agricultural producers along with realigning federal conservation programs to optimize pollinator forage and habitat (Breeze et al., 2014) can increase and diversify native bee populations to enhance global pollination security.

ACKNOWLEDGEMENTS

The authors first and foremost thank Amazon's Mechanical Turk survey respondents. We also thank Eric Venturini for collecting wildflower and clover bee pasture cost data, as well as Dr. Alison Dibble and Dr. Lois Stack for their input. This research was made possible through support from a U.S. Department of Agriculture, National Institute for Food and Agriculture, Specialty Crop Research Initiative Grant #2011-01389.

REFERENCES

Aizen MA, Harder LD (2009). The global stock of domesticated honey bees is growing slower than agricultural demand for pollination. Curr. Biol. 19: 915-918.

Allsopp MH, DeLange W, Veldtman R (2008). Valuing insect pollination services with cost of replacement. PLOS ONE 3(9): e3128. doi: 10.1371/journal.pone.0003128.

Asare E (2013). The economic impacts of bee pollination on the profitability of the lowbush blueberry industry in Maine. Master's thesis, The University of Maine, Orono, ME.

Baetschmann G (2012). Identification and estimation of thresholds in the fixed effects ordered logit model. Econ. Letters 115(3): 416-418.

Blaauw BR, Isaacs R (2014). Flower plantings increase wild bee abundance and the pollination services provided to a pollination-dependent crop. J. Appl. Ecol. 51(4): 890-898.

Breeze TD, Bailey AP, Balcombe KG, Potts SG (2014). Costing conservation: An expert appraisal of the pollinator habitat benefits of England's entry level stewardship. Biodiver. Conserv. 23(5) 1193-1214.

Bushmann SL, Drummond FA. In Press. Abundance and diversity of wild bees (Hymenoptera: Apoidea) found in lowbush blueberry growing regions of Downeast Maine. Environ. Entomol.

Carreck NL, Williams IH (2002). Food for insect pollinators on farmland: insect visits to flowers of annual seed mixtures. J. Insect Conserv. 6(1): 13-23.

Darby K, Batte MT, Ernst S, Roe B (2006). Willingness to pay for locally produced foods: A customer intercept study of direct market and grocery store shoppers. Am. Agric. Econ. Assoc., Sel. Papers: 1-31.

Drummond FA, Aronstein K, Chen J, Ellis J, Evans J, Ostiguy N, Sheppard W, Spivak M, Visscher K (2012). The First Two Years of the Stationary Hive Project: Abiotic Site Effects. Amer. Bee J. April 2012.

Gallai N, Salles J-M, Settele J, Vaissière BE (2009). Economic valuation of the vulnerability of world agriculture confronted with pollinator decline. Ecol. Econ. 68(3): 810-821.

Garibaldi LA, Stefan-Dewenter I, Winfree R, Aizen MA, Bommarco R, Cunningham SA, Kremen C, Carvalheiro LG, Harder LD, Afik O, Bartomeus I, Benjamin F, Boreux V, Cariveau D, Chacoff NP, Dudenhöffer JH, Freitas BM, Ghazoul J, Greenleaf S, Hipólito J, Holzschuh A, Howlett B, Isaacs R, Javorek SK, Kennedy CM, Krewenka KM, Krishnan S, Mandelik Y, Mayfield MM, Motzke I, Munyuli T, Nault BA, Otieno M, Petersen J, Pisanty G, Potts SG, Rader R, Ricketts TH, Rundlöf M, Seymour CL, Schüepp C, Szentgyörgyi H, Taki H, Tscharntke T, Vergara CH, Viana BF, Wanger TC, Westphal C, Williams N, Klein AM (2013). Wild pollinators enhance fruit set of crops regardless of honey bee abundance. Science 339: 1608-1611.

Garibaldi LA, Carvalheiro LG, Leonhardt SD, Aizen MA, Blaauw BR, Isaacs R, Kuhlmann M, Kleijn D, Klein AM, Kremen C, Morandin L, Scheper J, Winfree R (2014). From research to action: enhancing crop yield through wild pollinators. Front. in Ecol. Env. 12(8): 439-447.

Goulson D (2013). An overview of the environmental risks posed by neonicotinoid insecticides. J. Appl. Ecol. 50, 977–987. doi:10.1111/1365-2664.12111

Goulson D, Nicholls E, Botias C, Rotheray E (2015). Bee declines driven by combined stress from parasites, pesticides, and lack of flowers. Scienc express/ sciencemag.org/content/early/recent / 26 February 2015 / Page 1 / 10.1126/science.1255957

Greene WH (2012). Econometric Analysis. Pearson Publ., Boston, MA.

Greene WH, Hensher DA (2010). Modeling Ordered Choices. A Primer, Cambridge University Press, Cambridge, UK.

Haaland C, Naisbit RE, Bersier L-F (2011). Sown wildflower strips for insect conservation: a review. Insect Conserv. Diver. 4: 60-80.

Hanes SP, Collum K, Hoshide AK, Asare E (2013). Grower perceptions of native pollinators and pollination strategies in the lowbush blueberry industry. Renewable Agriculture and Food Systems 28 (4): 1-8.

Hu W, Woods T, Bastin S (2009). Consumer acceptance and willingness to pay for blueberry products with nonconventional attributes. J. Agric. and Appl. Econ. 41(1): 47-60.

Hu W, Woods T, Bastin S, Cox L, You W (2011). Assessing consumer willingness to pay for value-added blueberry products using a payment card survey. J. Agric. and Appl. Econ. 43(2), 243-258.

Isaacs R, Kirk AK (2010). Pollination services provided to small and large highbush blueberry fields by wild and managed bees. J. Appl. Ecol. 47(4): 841-849.

Jones MS, Vanhanen H, Peltola R, Drummond FA (2014). A global review of arthropod-mediated ecosystem-services in Vaccinium berry agroecosystems. Terrestrial Arthro. Rev. 7: 41-78.

Kay RD (2011). Farm Management: Planning, Control, and Implementation, Seventh Edition. New York: McGraw-Hill, Inc., 480 p.

Loureiro ML, McCluskey JJ, Mittelhammer RC (2001). Assessing consumer preferences for organic, eco-labeled, and regular apples. J. Agric. and Res. Econ. 26(2): 404-416.

Masori W, Suri S (2012). Conducting behavioral research on Amazon's Mechanical Turk. Behav. Res. 44: 1-23.

Murphy JJ, Stevens TH (2004). Contingent valuation and hypothetical bias in experimental economics. Agric. and Res. Econ. Rev. 33(2): 182-92.

NRC (National Resource Council) of the National Academies (2007). Status of pollinators in North America. Committee on the Status of Pollinators in North America, National Research Council of the National Academies, National Academies Press, Washington D.C.

Neumann P, Carreck NL (2010). Honey bee colony losses. J. Apicul. Res. 49(1): 1-6. DOI 10.3896/IBRA.1.49.1.01

Nicholls CI, Altieri MA (2013). Plant biodiversity enhances bees and other insect pollinators in agroecosystems. A review. Agron. Sustain. Dev. 33(2): 257-274.

Novotorova N (2007). A conjoint analysis of consumer preferences for product attributes: the case of Illinois apples. Doctoral dissertation, University of Illinois, Urbana, IL.

Olesen II, Alfnes FF, Rora MB, Kolstad KK (2010). Eliciting consumers' willingness to pay for organic and welfare-labeled salmon in a non-hypothetical choice experiment. Livestock Sci. 127(2-3): 218-226.

Ratnieks FW, Carreck NL (2010). Clarity on honey bee collapse? Science 327(5962): 152-153.

Rose A, Drummond FA, Yarborough DE, Asare E (2013. Maine wild blueberry growers: A 2010 economic and sociological analysis of a traditional Downeast crop in transition. Maine Agric. and Forest Exper. Stn. Misc. Rept. 445.

Rucker RR, Thurman WN, Burgett M (2012). Honey bee pollination markets and the internalization of reciprocal benefits. Am. J. Agric. Econ. 94(4): 936-977.

Scheper J, Reemer M, van Kats R, Ozinga WA, van der Linden GTJ, Schaminée JHJ, Siepel H, Kleijn D (2014). Museum specimens reveal loss of pollen host plants as key factor driving wild bee decline in The Netherlands. Proc. Nat. Acad. Sci. 111(49): 17,552-17,557.

Shi L, Gao Z, House L (2011). Consumer WTP for blueberry attributes: A hierarchical Bayesian approach in the WTP space. Agric. and Appl. Econ. Assoc., Sel. Paper.

Southwick EE, Southwick L (1992). Estimating the economic value of honey bees (Hymenoptera: Apidae) as agricultural pollinators in the United States. J. Econ. Ent. 85: 621-33.

EPA, NASS (U.S. Department of Agriculture, National Agricultural Statistics Service) (2010-2013). Statistics by State. NASS, Agricultural Statistics Board, USDA. http://www.nass.usda.gov. Accessed July, 29, 2015.

USDA, ARS (United States Department of Agriculture: Agricultural Research Service) (2012). Questions and Answers: Colony Collapse Disorder. http://www.ars.usda.gov/News/docs.htm?docid=15572. Accessed July, 29, 2015.

USEPA (United States Environmental Protection Agency) (2012). Colony Collapse Disorder: European Bans on Neonicotinoid Pesticides. http://www.epa.gov/pesticides/about/intheworks/ccd-european-ban.html. Accessed July, 29, 2015.

vanEngelsdorp D, Evans JD, Saegerman C, Mullin C, Haubruge E, Nguyen BK, Frazier M, Frazier J, Cox-Foster D, Chen Y, Underwood R, Tarpy DR, Pettis JS (2009). Colony Collapse Disorder: A Descriptive Study. PLOS ONE 4(8): e6481. Doi:10.1371/journal.pone.0006481

vanEngelsdorp D, Meixner M (2010). A historical review of managed honey bee populations in Europe and the United States and the factors that may affect them. J. of Invert. Path. 103: 580-595.

Winfree R, Gross BJ, Kremen C (2011). Valuing pollination services to agriculture. Ecol. Econ. 71: 80-88.

Wratten SD, Gillespie M, Decourtye A, Mader E, Desneux N (2012). Pollinator habitat enhancement: Benefits to other ecosystem services. Agric. Ecosys. Env. 159: 112-122.

Permissions

All chapters in this book were first published in IJAM, by Premier Publishers; hereby published with permission under the Creative Commons Attribution License or equivalent. Every chapter published in this book has been scrutinized by our experts. Their significance has been extensively debated. The topics covered herein carry significant findings which will fuel the growth of the discipline. They may even be implemented as practical applications or may be referred to as a beginning point for another development.

The contributors of this book come from diverse backgrounds, making this book a truly international effort. This book will bring forth new frontiers with its revolutionizing research information and detailed analysis of the nascent developments around the world.

We would like to thank all the contributing authors for lending their expertise to make the book truly unique. They have played a crucial role in the development of this book. Without their invaluable contributions this book wouldn't have been possible. They have made vital efforts to compile up to date information on the varied aspects of this subject to make this book a valuable addition to the collection of many professionals and students.

This book was conceptualized with the vision of imparting up-to-date information and advanced data in this field. To ensure the same, a matchless editorial board was set up. Every individual on the board went through rigorous rounds of assessment to prove their worth. After which they invested a large part of their time researching and compiling the most relevant data for our readers.

The editorial board has been involved in producing this book since its inception. They have spent rigorous hours researching and exploring the diverse topics which have resulted in the successful publishing of this book. They have passed on their knowledge of decades through this book. To expedite this challenging task, the publisher supported the team at every step. A small team of assistant editors was also appointed to further simplify the editing procedure and attain best results for the readers.

Apart from the editorial board, the designing team has also invested a significant amount of their time in understanding the subject and creating the most relevant covers. They scrutinized every image to scout for the most suitable representation of the subject and create an appropriate cover for the book.

The publishing team has been an ardent support to the editorial, designing and production team. Their endless efforts to recruit the best for this project, has resulted in the accomplishment of this book. They are a veteran in the field of academics and their pool of knowledge is as vast as their experience in printing. Their expertise and guidance has proved useful at every step. Their uncompromising quality standards have made this book an exceptional effort. Their encouragement from time to time has been an inspiration for everyone.

The publisher and the editorial board hope that this book will prove to be a valuable piece of knowledge for researchers, students, practitioners and scholars across the globe.

List of Contributors

Abel Leonard and D. M. Gabagambi
Department of Agricultural Economics and Agribusiness, Sokoine University of Agriculture, Morogoro, Tanzania

E. K. Batamuzi and E. D. Karimuribo
Faculty of Veterinary Medicine, Sokoine University of Agriculture, Morogoro, Tanzania

R. M. Wambura
Institute of Continuing Education, Sokoine University of Agriculture, Morogoro, Tanzania

Sobhy M. Ismaiel, Safar H. Al-Kahtani and Ali I. Saad
Department of Agriculture Economics, College of Agricultural Sciences, King Saud University, Riyadh- Saudi Arabia

Ahmed M. Al-Abdulkader
National Plan for Science, Technology and Innovation, King Abdulaziz City for Science and Technology, Riyadh- Saudi Arabia

Thaneshwar Bhandari
Agricultural Economics, Department of Agricultural Economics, Tribhuvan University- Institute of Agriculture and Animal Science, Lamjung, Nepal

Oluwatuyi Toyin Bukola, Eronmwon Iyore and Emokpae Osayi Precious
Department of Agricultural Economics and Extension services, Faculty of Agriculture, University of Benin, Benin City, Nigeria

K. C. Dilli Bahadur, N. Gadal, S. P. Neupane, R. R. Puri, B. Khatiwada, G. Ortiz Ferrara, A. R. Sadananda and C. Böber
International Maize and Wheat Improvement Center (CIMMYT Intl., Mexico), CGIAR Center, Nepal

Rabirou Kassali
Department of Agricultural Economics, Obafemi Awolowo University, Ile-Ife, Osun State, Nigeria

Abdulhameed Abana Girei and Ismaila Dauda Sanu
Department of Agricultural Economics and Extension, Nasarawa State University, Keffi, Nigeria

Joseph Gichuru Wang'ombe
African Population and Health Research Center, Nairobi 00100, Kenya

Meine Pieter van Dijk
Maastricht School of Management, Maastricht, Netherlands
UNESCO-IHE Institute for Water education, Delft, Netherlands

Tesfaye Berihun
College of Business and Economics, Hawassa University, Hawassa, Ethiopia

Mansoor Ahmed Koondhar, Abbas Ali Chandio and He Ge
College of Economics and Management, Sichuan Agriculture University, Chengdu, 611130 China

Mumtaz Ali Joyo, Masroor Ali Koondhar and Riaz Hussain Jamali
Department of Agricultural Economics, Sindh Agriculture University Tandojam, Pakistan

Ravi Nandi
Program Manager, National Institute of Agricultural Extension Management (MANAGE), Hyderabad, 500030 Rajendranagar, Telengana State, India

Nithya Vishwanath Gowdru
National Institute of Rural Development and Panchayat Raj (NIRD&PR), Hyderabad, 500030 Rajendranagar, Telengana State, India

Wolfgang Bokelmann
Department of Agricultural and Horticultural Economics, Invalidenstasse 42, The Humboldt University of Berlin, Germany

S. S. R. M. Mahe Alam Sorwar, Md. Tanvir Ahmed, Sudhir Chandra Nath and Md. Harun -OR- Rashid
Seed and Agro Enterprise, BRAC, Bangladesh

Chris Wheatley
International Potato Centre, Philippines

Francis Lwesya and Vicent Kibambila
Department of Business Administration, School of Business Studies and Economics, University of Dodoma, Tanzania

Chala Hailu and Chalchisa Fana
Department of Agribusiness and Value Chain Management, College of Agriculture and Veterinary Science, Ambo University, Ethiopia

Benigno Rodríguez-Padróni
Chapingo Autonomous University, Chapingo, State of Mexico

André Ricardo Cortés Jarrín and Kees Burger
Development Economics Group, Wageningen School of Social Sciences, Wageningen University, Hollandseweg, The Netherlands

S. H. Alkahtani and S. M. Ismaiel
King Saud University, Saudi Arabia

A. M. Al-Abdulkader
King Abdulaziz City for Science and Technology, Riyadh 11451, Saudi Arabia

Ben-Chendo G. N, N. Lawal, M. N. Osuji, I. I. Osugiri and B. O. Ibeagwa
Department of Agricultural Economics, Federal University of Technology Owerri, Imo State, Nigeria

Thomas Stevens
Department of Resource Economics, 224 Stockbridge Hall, University of Massachusetts, Amherst, MA 01003, USA

Aaron K. Hoshide
School of Economics, 206 Winslow Hall, University of Maine, Orono, ME 04469, USA

Francis A. Drummond
School of Biology and Ecology, 305 Deering Hall, University of Maine, Orono, ME 04469, USA
University of Maine Cooperative Extension, 5722 Deering Hall, University of Maine, Orono, ME 04469, USA

Index

www.ingramcontent.com/pod-product-compliance
Lightning Source LLC
Chambersburg PA
CBHW080300230326
41458CB00097B/5247